New Radio Laboratory of the Bureau of Standards (under construction February, 1918)

DEPARTMENT OF COMMERCE

CIRCULAR

OF THE

U.S. BUREAU OF STANDARDS

S. W. STRATTON, Director

No. 74

RADIO INSTRUMENTS AND MEASUREMENTS

ISSUED MARCH 23, 1918

PRICE, 60 CENTS

Sold only by the Superintendent of Documents, Government Printing Office
Washington, D. C.

WASHINGTON
GOVERNMENT PRINTING OFFICE
1918

CONTENTS

3

4   *Contents*

RADIO INSTRUMENTS AND MEASUREMENTS

PART I.—THEORETICAL BASIS OF RADIO MEASUREMENTS

INTRODUCTION

In the rapid growth of radio communication, the appliances and methods used have undergone frequent and radical changes. In this growth, progress has been made largely by new inventions and by the use of greater power, and comparatively little attention paid to refinements of measurement. In consequence the methods and instruments of measurement peculiar to radio science have developed slowly and have not yet been carried to a point where they are as accurate or as well standardized as other electrical measurements.

This circular presents information regarding the more important instruments and measurements actually used in radio work. It is hoped that the treatment will be of interest and value to Government officers, radio engineers, and others, notwithstanding the subject is not completely covered. Many of the matters dealt with are or have been under investigation in the laboratories of this Bureau and are not treated in previously existing publications. No attempt is made in this circular to deal with the operation of apparatus in sending and receiving. It is hoped to deal with such apparatus in a future circular. The present circular will be revised from time to time, in order to supplement the information given and to keep pace with progress. The Bureau will greatly appreciate suggestions from those who use the publication for improvements or changes which would make it more useful in military or other service.

The methods, formulas, and data used in radio work can not be properly understood or effectively used without a knowledge of the principles on which they are based. The first part of this circular, therefore, attempts to give a summary of these principles

5

in a form that is as simple as is consistent with accuracy. A large proportion of this publication is devoted to the treatment of fundamental principles for the reasons, first, that however much the methods and technique of radio measurement may change the same principles continue to apply, and second, that this will make the present circular serve better as an introduction to other circulars on radio subjects which may be issued.

A familiarity with elementary electrical theory and practice is assumed. Introductory treatment of electrostatics and magnetic poles, electric and magnetic fields, the laws of direct currents, and descriptions of the more common electric instruments and experiments may be found in many books. A list of publications suitable as an introduction to the theory given in this circular may be found in the bibliography (p. 324). The common explanation of electric current as similar to the flow of water in a pipe, while adequate for most of the phenomena of direct current is not suitable for alternating currents and particularly for radio. The explanations here given attempt to give a better insight into the behavior of electric current. Most of the treatment of principles is a presentation of the theory of low-frequency alternating currents, arranged with its radio applications in mind. There is little in the way of special theory before section 24, which deals with damped waves, and yet the underlying principles of the chief radio phenomena are covered. Furthermore, damped waves are of less importance than formerly, since modern practice tends toward the exclusive use of continuous or undamped waves. The principles of radio measurements are thus nearly identical with those of any other alternating-current measurements.

THE FUNDAMENTALS OF ELECTROMAGNETISM

1. ELECTRIC CURRENT

Electric current is the rate of flow of a quantity of electricity. The most familiar and most useful properties of an electric current are (1) the heating effect produced in a conductor in which it flows, and (2) the magnetic field surrounding it. The latter is by far the most important property in radio work. The study of the combined effects of electricity and the magnetism accompanying an electric current constitutes the subject of electromagnetism.

When a current flows continuously in the same direction, as the current from a battery, it is called a direct current. When the current periodically reverses in direction, it is an alternating current. The alternation of current is accompanied by a reversal of direction of the magnetic field around the current. On this account alternating currents behave very differently from direct currents. The uses of alternating currents may be divided, roughly, into three groups, separated according to the frequency of alternation of the currents used:

Electric power applications, 20 to 100 per second.

Telephony, 100 to 20 000 per second.

Radio, 20 000 to 2 000 000 per second.

Displacement Currents.—Direct currents can flow continuously in conductors only, while alternating currents flow also in insulators. Suppose a circuit contains a condenser, consisting of two large metal plates separated by air or some other insulating medium. If a battery is connected into the circuit, a momentary flow of current takes place, accompanied by an electric strain in the insulating medium of the condenser. This strain is opposed by an electric stress, which soon stops the current flow. The action is much like the flow of gas under pressure into a gas tank, as described below in the section on capacity. The flow of gas stops as soon as the back pressure of the compressed gas in the tank is equal to the applied pressure. The flow of electric current stops as soon as the back electric pressure (called "potential difference") of the electrically strained medium is equal to the electric pressure ("electromotive force") of the battery.

If there is a source of alternating current in a circuit containing a condenser, the electric strain in the insulating medium reverses in direction for every alternation of the current. The electric strain of which we have been speaking is called electric displacement and its variation gives rise to a so-called displacement current. The electric strain is of two kinds: First, there is the strain in the actual material dielectric. This part is a movement of electricity in the same sense as the transference of a definite quantity of electricity through a wire is a movement of electricity, the only difference being that in the insulator there is a force (which we called electric stress) which acts against the electric displacement. Second, there is the strain which would exist in the ether if the material dielectric were absent. Some prefer to think of this as a sort of electric displacement in the ether, of a kind similar to that in the matter; but a knowledge of the physical nature of electric displacement is unnecessary for practical purposes; all that is necessary is a statement of how the displacement current is to be defined as measured, and of the relation of the quantity so measured to other electromagnetic quantities.

The effect of electric displacement at any point in a medium is handed on to adjacent points and so spreads out through space. Under certain conditions a considerable quantity of this moving displacement and the magnetic field accompanying it become detached from the circuit. This process is what constitutes the radiation of electromagnetic waves, which makes radio communication possible.

Electrons.—The flow of an electric current in a conductor is not opposed by electric stress as in an insulator. A current in a conductor is believed to consist of the motion of immense numbers of extremely small particles of electricity, called electrons. All electrons are, so far as known, strictly identical, are, for ease of calculation, assumed to be spherical in shape, and have the following dimensions, etc.: Radius, 1×10^{-13} centimeter; mass, 8.8×10^{-28} gram; electric charge, 1.59×10^{-19} coulomb.

An electron is thousands of times smaller than any atom. Views as to the transference of the electric current by the motion of electrons in a conductor have undergone considerable changes during the last few years. Some of the electrons in a conductor are bound to the molecules while others are free to move about. The latter are in constant motion between the molecules, in zigzag paths because of repeated collisions with the molecules. The motion of the free electrons is thus very similar to the heat motions of the molecules; in fact, the average kinetic energy of an electron is equal to that of a molecule at the same temperature. The average velocity of the electrons is about 100 km per second at 0° C. This increases with temperature. Until recently it was supposed that conduction in a solid takes place almost entirely through the agency of the free electrons. The electric current was nothing more than a slow drift (a sort of electric wind) superposed upon the random motions of the electrons by the electric field. This view, although suggestive and fruitful, is attended with many difficulties, and the present tendency is toward some form of theory in which the conduction is brought about by the spontaneous discharge of electrons from one molecule to another, the function of the field being to influence the orientation of the discharge, which would, in its absence, be perfectly random. The study of electrons has recently led to great improvements in apparatus for producing and detecting currents of radio frequency. (See sec. 56.)

When electricity is in equilibrium in a conductor, the electric charge is in a very thin layer upon its surface; thus the phenomena of electrostatics arise. An insulator is believed to contain no free electrons. The electrons are bound to the molecules in such a way that they can be slightly displaced by an electric force but return to their positions of equilibrium when it is removed. This motion of the bound electrons, together with the electric strain in the ether itself, constitutes the electric displacement in the insulator, and determines its dielectric constant. Another type of motion of which the electrons attached to molecules are capable is vibration about their positions of equilibrium. They thus give rise to waves that travel outward in the form of light and heat radiation.

2. ENERGY

Most useful operations in physics involve the movement of something from one place to another; and in general, as for example when the body moved is held by a spring, or when its velocity changes during the motion, force has to be exerted to cause the motion. A useful quantity which figures in the discussion of such motions is the quantity called "work." The work done upon a body is defined as the product of the force which acts upon the body into the distance moved; or (when the force varies during the motion) as the sum of such products for each element of path described by the body. When a body or system of any kind possesses the power to do work in virtue of its position, velocity, chemical constitution, or any other feature, it is said to possess energy; and the measure of the change of energy which it experiences in doing such work is the amount of work done. One of the fundamental principles of science is the "conservation of energy." The amount of energy in existence is constant; energy can not be created nor destroyed; it can only be transformed from one form into another. It is often very helpful to the understanding of a process to consider what energy changes are taking place. The transformation from one form into another is always accompanied by a dissipation of some of the energy as heat or some other form in which it is no longer available for the use desired. Thus while none of the energy is lost during an energy change, more or less of it becomes no longer available.

Kinds of Energy.—The familiar kinds of energy are mechanical energy, heat, chemical energy and electrical energy. To these may be added radiant energy, but this is considered to be a form of electrical energy. Mechanical energy is of two kinds, kinetic energy and potential energy. When an object is in motion it is said to possess kinetic energy. If the motion is stopped the kinetic energy of the object changes into some other form. For example, if the moving object is stopped by suddenly striking an immovable obstacle its kinetic energy is converted into heat. If when stopping it starts another object moving, the second object then has kinetic energy. Any object of mass m moving with velocity v has a kinetic energy $= \frac{1}{2} m v^2$.

The energy which an object possesses in virtue of its position is called potential energy. A stone, lifted a certain distance above the earth, will fall if released. It then acquires kinetic energy in falling. It had a certain amount of potential energy when at the highest point, simply in virtue of its position above the earth. As it falls this potential energy is being changed into kinetic energy, and when it is just about to strike the ground the potential energy has all been converted into kinetic. This is a simple example of the principle of conservation of energy.

When a change of energy from one form into another occurs, work is done. When an object falls to the earth from a height there is a change of potential energy into kinetic and work is done upon the object by the force of gravity, the amount of which is equal to the product of the force by the distance through which the object falls. Again, when a body is moved against a force tending to oppose the motion, work must be done by the agency which moves the body. The product of these two factors, the force acting and the displacement of the object, is the amount of work done in an energy transformation.

Electrical Energy.—There are two kinds of electrical energy, similar to the two kinds of mechanical energy. Corresponding to potential energy there is electrostatic energy, which is the energy of position of electricity at rest; this is the form in which electricity is stored in a charged' condenser. Corresponding to kinetic energy there is electrokinetic energy (also called magnetic energy), which is the energy of electricity in motion. The latter is the energy of the electric current, and is associated with the magnetic field accompanying the current. In accordance with the law of conservation of energy the sum of the electrostatic and the magnetic energies in any electrical system is constant if the system as a whole does not receive or give out any energy, or if energy is being supplied at the same rate at which it is being dissipated.

Electrical energy can readily be converted into other types of energy; if this were not so it would not be the important factor in the life of man that it now is. As far as radio science is concerned, the two principal forms of energy into which electrical energy is transformed are heat and electromagnetic radiation. From any electrical system there is a continuous dissipation or loss of electrical energy going on, and in general the evolution of heat in the circuit plus the energy radiated as electromagnetic waves equals the diminution of the sum of the electrostatic and magnetic energies. The energy of electromagnetic waves is a form of electric energy, being a combination of electrostatic and magnetic energies. Inasmuch, however, as it travels through space entirely detached from the sending circuit, it represents a loss of energy from that circuit.

3. RESISTANCE

The dissipation or loss of electrical energy is expressible in terms of resistance. The rate of evolution of energy at any instant in a conductor is the product of the electromotive force acting in the conductor by the current flowing. This energy usually manifests itself in the form of heat. The time rate of energy is called power. Thus,

$$p = \frac{w}{t} = e\,i = R\,i^2 = \frac{e^2}{R}$$

Resistance is defined by $R = \frac{e}{i}$. Power (p) is generally expressed in watts, energy (w) in joules, electromotive force (e) in volts, current (i) in amperes, and resistance (R) in ohms, unless otherwise stated. The resistance of a conductor depends on the material of which it is made, the size and shape of the conductor, and the frequency of the current. The characteristic property of the material is called its resistivity. Denoting by ρ the ordinary or volume resistivity, by l the length, by S the cross section of the conductor, and by R_0 the resistance to direct current,

$$R_0 = \rho \frac{l}{S}$$

The resistance of a system of conductors is readily found by the simple laws of series and parallel combination, for direct currents.

With alternating currents, however, the calculation is more difficult, and it is usually found convenient to utilize the relation

$$R = \frac{p}{i^2}$$ (1)

The resistance of a single conductor to alternating currents is found by the aid of the same relation. It may be shown that the distribution of direct current in a system of conductors or over the cross section of a single conductor is such as to make the production of heat a minimum; and it results that in a single uniform conductor the current is uniformly distributed over the cross section. When alternating current flows in a conductor, it tends to flow more in the outer portions of the conductor than in the center. In consequence of this change of current distribution, the power which is converted into heat increases. The higher the frequency the farther does the current distribution depart from the direct-current distribution, and the greater does the power p become. It follows, in accordance with equation (1), that the resistance increases as frequency increases.

Radio-Frequency Resistance.—With alternating currents the departure from uniform distribution of the current is spoken of as the skin effect. At high frequencies the current flows in a thin layer at the surfaces of conductors, and the skin effect is thus large in all except very thin wires; the resistances of ordinary conductors at radio frequencies may be many times their low-frequency resistances. The ratio of resistance at any frequency to the direct-current or low-frequency resistance can be calculated for certain simple forms of conductors. Formulas for this are given below in sections 74 to 76. In most practical cases, however, the radio resistance can be obtained only by measurement.

In addition to the resistances of conductors, resistance is introduced into radio circuits by three other causes, viz, sparks, dielectrics, and radiation. Dielectric resistance is treated in section 34, below. The energy radiated from a circuit per unit time in electromagnetic waves is proportional to the square of the current in the circuit. It is thus analogous to the energy dissipation as heat in a conductor, and, therefore, the radiation increases the effective or equivalent resistance by a certain amount. This added resistance is conveniently called radiation resistance. It can be calculated for a few simple types of circuit. It is, in general, large enough to be appreciable only when the circuit has the open or antenna form, or when a closed circuit is of large area and the frequency is high.

4. CAPACITY

Electrostatic energy may be stored in an arrangement of conductors and insulator called an electrical condenser. The action of a condenser is somewhat similar to that of a gas tank used for the storage of gas. The amount of gas a tank will hold is not a constant, fixed amount; it depends on the pressure. If the pressure is doubled, twice the mass of gas is forced into the tank. The internal or back pressure of the gas opposes the applied pressure. If the applied pressure is released and an opening is left in the tank, the gas rushes forth.

The amount of electric charge given to a condenser depends on the electric pressure, or potential difference; and in exact similarity to the gas, the charge is proportional to this potential difference. The constant ratio of charge to potential difference is called the capacity of the condenser. In symbols, $\frac{Q}{V} = C$. The capacity of a condenser depends on the size and distance apart of its plates, and on the kind of dielectric between the plates. (Various kinds of condensers are described in sec. 32, etc., and formulas for calculating capacity are given in secs. 63 to 65.) The applied potential difference is opposed by a sort of elastic reaction of the electricity in the condenser, just as the internal pressure of the gas in a tank opposes the external applied pressure. If the plates of a charged condenser are connected by a conductor, with no applied electromotive force, the condenser discharges.

The Dielectric.—The insulating medium in a condenser is called a dielectric. The process of charging causes electric displacement in the dielectric. When a body is moved against a force tending to prevent the motion, work is done, and similarly, when a condenser is charged against the quasi-elastic reaction of the dielectric, work is done upon the condenser. The energy of the charging source is stored up as electrostatic energy in the dielectric. The two factors upon which the energy depends are the charge and the potential difference,

$$W = \frac{1}{2}QV$$

Since $\frac{Q}{V} = C$, we have also

$$W = \frac{1}{2}V^2C = \frac{Q^2}{2C}$$

The pressure upon the gas in a tank can not be increased indefinitely, for the tank will ultimately yield and break. Similarly there is a limit to the potential difference which can be applied to a condenser, for the dielectric will be broken down, or punctured, if the limit is exceeded. The potential difference at which a spark will pass and the dielectric be punctured is called the " dielectric strength."

Capacity is one of the two quantities of chief importance in radio circuits. The other is inductance, treated in the following section.

5. INDUCTANCE

Magnetic Flux.—The physical quantity called inductance is dependent upon the magnetic field which surrounds every electric current. The intensity of this magnetic field at any point is proportional to the c u r r e n t. The direction of the magnetic field around a straight wire carrying a current is given by Ampere's right-hand rule: Close the right hand with the thumb extended; point the thumb in the direction of the current flow; the magnetic field is then in the direction in which the fingers point, in circles in planes perpendicular to the wire. The magnitude of the magnetic field intensity can be easily computed for some simple forms of circuit from the principle that its line integral [1] in a path completely around the current is equal to $\frac{4\pi}{10}$ times the current in amperes.

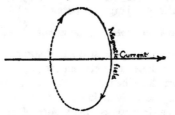

FIG. 1.—*Direction of magnetic field around a wire carrying a current*

As an example, suppose a current i flowing in a very long solenoid of N turns, radius r and length l. The magnetic field intensity H is parallel to the axis and of constant value inside the solenoid; it may be shown to be zero in the space outside, and the effects of the ends may be neglected. The line integral of H along any path completely around the current is Hl inside the solenoid

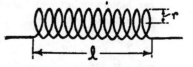

FIG. 2.—*Simple solenoid*

---

[1] The line integral of a quantity along any line or path is the sum of the products of the length of each element of the path by the value of the quantity along that element. If the quantity has a constant value along the whole path, the line integral is simply the product of this value by the length of the path.

and is zero for the rest of the path. The path incloses the current i, N times. Hence

$$Hl = \frac{4\pi}{10} Ni, \text{ or}$$

$$H = \frac{4\pi}{10} \frac{Ni}{l} \tag{2}$$

The magnetic field in the medium surrounding a conductor carrying a current produces a magnetized condition of the medium. This condition is a sort of magnetic strain in the medium and is analogous to displacement produced in a dielectric by electric potential difference. The amount of this magnetic strain through any area is called the magnetic flux. This quantity (for which the symbol ϕ is used) is equal to the product of the three factors, magnetic field intensity, area and the magnetic permeability. Permeability is a property of matter or of any medium which indicates, so to speak, its magnetizability. Its numerical value is equal to unity for empty space and for air and most substances. Iron may have a permeability as high as 10 000 or even more.

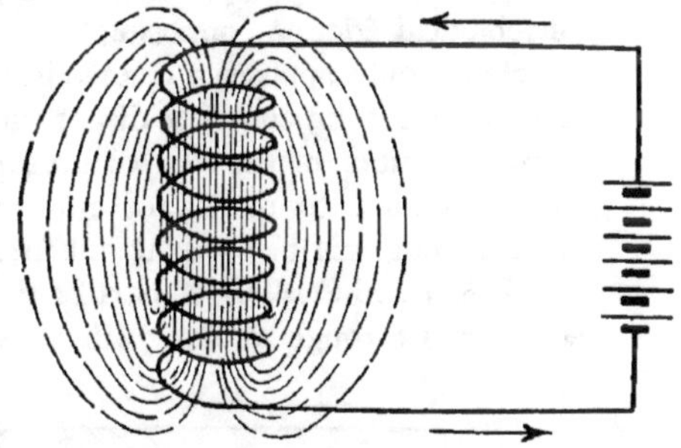

Self - inductance.— Inductance is a quantity introduced as a

FIG. 3.—*Magnetic flux around a solenoid in which a current is flowing.*

convenient means of dealing with magnetic fluxes associated with currents. The self-inductance of a circuit is simply the total magnetic flux linked with the circuit due to a current in the circuit, per unit of current. In symbols,

$$L = \frac{\phi}{i} \tag{3}$$

The magnitude of L depends on the shape and size of the circuit and is a constant for a given circuit, the surrounding medium being of constant permeability. If the circuit has N turns each traversed by the same magnetic flux ϕ, then when L is expressed in terms of the usual unit, called the "henry,"

$$L = \frac{N\phi}{10^8 i} \tag{4}$$

Analogy of Inductance to Inertia.—The magnetic flux associated with a current is analogous to the momentum associated
with a moving body. Because of its inertia or mass (m), a body
in motion with velocity (v) opposes any change in its momentum
(mv). Inertia is obviously a very different thing from friction,
which always resists the motion and tends to decrease the velocity
or momentum. Inertia only opposes a change in momentum,

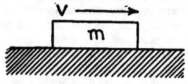

and hence does not affect a
motion with constant velocity.
Inductance may be spoken of
as electrical inertia or mass.
The inductance (L) of a circuit in which a current (i) is
flowing opposes any change in
the flux (ϕ). Electrical resist-

Fig. 4.—*A body having a mass (m) and a velocity
(v) opposes any change in its momentum (mv)*

ance and inductance are very different, for resistance behaves like
mechanical friction, opposing even a constant current, while inductance only opposes a change in the current. Thus inductance
has no effect on constant direct currents but is one of the determining factors in the flow of alternating currents. The analogy
of inductance to inertia, of current to velocity, and of flux to
momentum, will be brought out further in the next section.

Mutual Inductance.—A part of the magnetic flux from a circuit
may pass through or link with a second circuit. The amount of

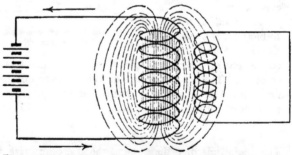

Fig. 5.—*Linking of magnetic flux of one circuit with another; the basis of
the conception of mutual inductance*

this flux linked with circuit 2, per unit of current in circuit 1, is
called the mutual inductance of the two circuits. If ϕ_{12} denotes
the flux mentioned and $i_1 =$ current in circuit 1, the mutual
inductance =

$$M = \frac{\phi_{12}}{i_1}$$

It is also true that if ϕ_{21} = the flux from circuit 2 linked with circuit 1, and i_2 = current in circuit 2,

$$M = \frac{\phi_{21}}{i_2}$$

The magnitude of any mutual inductance depends on the shape and size of the two circuits, their positions and distance apart, and the permeability of the medium. If there are N_1 turns in the first circuit and N_2 turns in the second and the same amount of flux from one passes through every turn of the other, then using the ϕ's to denote that part of the flux from one turn of either circuit which passes through each turn of the other circuit, and using the usual units,

$$M = \frac{N_1 N_2 \phi_{12}}{10^8 i_1}$$

If, however, the ϕ's denote the total flux from either circuit passing through each turn of the other circuit, this becomes

$$M = \frac{N_2 \phi_{12}}{10^8 i_1} \tag{5}$$

For any of the definitions of ϕ, the ratio $\frac{\phi_{21}}{i_2} = \frac{\phi_{10}}{i_1}$, and is a quantity depending only on geometrical configuration.

Calculation of Inductance.—It is frequently convenient to deal with the self-inductance of a particular coil or with the mutual inductance of limited portions of two circuits. Inductance is strictly defined only for complete circuits. The self-inductance of a part of a circuit is understood to be such that the inductance of the complete circuit is equal to the sum of the self-inductances of all the parts and the mutual inductance of every part with every other part.

Inductances are computed by the aid of equation (3), together with the principle given above that flux is field intensity times area times permeability. For example, to find the inductance of the long solenoid of Fig. 2,

$$L = \frac{N}{10^8 i} \times H \times S \times \mu$$

in which $S=$ the area of the circular cross section of the solenoid and $\mu=1$, the permeability of air. Substituting the value of H from (2), and putting $S=\pi r^2$,

$$L = \frac{N}{10^8 i} \times \frac{4\pi}{10}\frac{Ni}{l} \times \pi r^2 \times 1 = \frac{4\pi^2}{10^9}\frac{N^2 r^2}{l}$$

Change of Inductance with Frequency.—When it is desired to calculate inductance with great accuracy, account must be taken of the magnetic flux within the conductor carrying the current, as well as the flux outside the conductor. The flux in a wire is greatest at the circumference and zero at the axis of the wire, because the flux is due only to the current which it surrounds. Any change in the distribution of current within the wire changes also the flux distribution and hence the inductance. As has been stated in the section on resistance, the current distribution is different for different frequencies. Consequently inductance varies with frequency. As the frequency is increased, less current flows near the axis of the wire and more flows in the surface portions. The flux in the central parts of the wire is thus diminished, and the inductance decreases as frequency increases. This effect is small, because the whole flux within the wire is a small part of the total flux. (See formulas (131) to (138) in sec. 67.) There is a similar change of mutual inductance with frequency, but it is so small as to be wholly negligible.

Series and Parallel Arrangement of Inductances.—Inductances in series add like resistances. When the coils or conductors which are combined are so far apart that mutual inductances are negligible, inductances in parallel are combined like resistances in parallel. Taking account of mutual inductance, the total inductance of any number of inductances in *series* is

$$L = L_1 + L_2 + L_3 + \ldots + 2(M_{12} + M_{13} + M_{23} + \ldots)$$

Some or all of the mutual inductances may be negative. For two coils in *parallel*, the total inductance is

$$L = \frac{L_1 L_2 - M^2}{L_1 + L_2 - 2M}$$

The last term in the denominator changes sign if the coils are so connected that M is negative. This expression applies at radio frequencies, but at low frequencies the resistance of the coil may have to be taken into account. For more than two inductances in parallel, the expression for the total inductance is complicated.

THE PRINCIPLES OF ALTERNATING CURRENTS

6. INDUCED ELECTROMOTIVE FORCE

When the magnetic flux through any circuit is changing, an electromotive force is produced around the circuit, which lasts while the change is going on. The change of flux may be caused in various ways; a magnet may be moved in the vicinity, the circuit or a part of it may be moved while near a magnet, the current in a second near-by circuit may be altered, or either circuit or a part thereof may be moved. The electromotive force thus caused is called an induced electromotive force, and the resulting current in the circuit is an induced current. The direction of the induced emf and current is given by Lenz's law, viz, an induced current always flows in such a direction as to oppose the action which produces it. For example, if the current is induced by bringing a magnet near a circuit, the current in the circuit will be such as to repel the magnet. The energy of the induced current results from the work necessary to bring up the magnet against the repelling force.

The magnitude of the induced emf at any instant is in every case equal to the rate of change of the magnetic flux through the circuit. This is expressed by the formula

$$e = \frac{d\phi}{dt} \tag{6}$$

where e represents the instantaneous value of the induced emf in a circuit consisting of a simple loop or single turn, and $\frac{d\phi}{dt}$ is an expression called the derivative [2] of flux with respect to time and which tells the instantaneous rate of change of the flux. This quantity $\frac{d\phi}{dt}$ is the change of flux during a very small interval of time divided by the time, and its value may vary from instant to instant. If it remains constant for a certain length of time, then its value is the whole change of flux during that interval divided by the interval.

If the changing flux through the circuit is the flux ϕ_{21} from a second circuit, $e = \frac{d\phi_{21}}{dt}$. This flux is expressible in terms of mu-

---

[2] While derivatives are used in a few places in this circular, it is believed that the treatment can, nevertheless, be understood by a person not familiar with calculus. To avoid the use of derivatives entirely would require circumlocution such that the treatment would doubtless be even less clear.

tual inductance and the current in the second circuit, thus, $\phi_{21} = Mi_2$. Consequently,

$$e = \frac{d(Mi_2)}{dt}.$$

If the circuits are fixed in position, M remains constant, so this becomes

$$e = M\frac{di_2}{dt} \qquad (7)$$

Thus the electromotive force induced in a circuit by variation of current in another circuit is equal to the product of the mutual inductance by the rate of change of current in the second circuit. An emf may also be induced in one circuit by a variation of the current in the circuit itself. Since the flux associated with a current is $\phi = Li$, it follows that the self-induced emf is given by

$$e = L\frac{di}{dt} \qquad (8)$$

That is, the emf induced in the circuit is equal to the product of the self-inductance by the rate of change of the current.

When the flux ϕ through a coil of N turns is changing, the total emf e induced in the whole coil is N times that induced in one turn. The simple equation (6) becomes, in terms of the usual units,

$$e = \frac{N}{10^8}\frac{d\phi}{dt} \qquad (9)$$

Equations (7) and (8) are correct when emf is expressed in volts, inductance in henries, current in amperes, and time in seconds. They were obtained from the simpler equation (6) for induced emf, but they need no modification on this account, because equations (4) and (5) for self and mutual inductance also contain the factors N and 10^8 which cancel those in (9).

Mechanical Analog.—The fact that a change of magnetic flux gives rise to an electromotive force which opposes the change may be understood by recalling that a change of mechanical momentum of a body is opposed by the force of inertia. This force is equal to the rate of change of momentum. The electrical and mechanical cases are strictly analogous. Flux corresponds to momentum, electromotive force to mechanical force (F), current to velocity (v), and inductance to mass (m). If the mass is constant, we have

$$F = m\frac{dv}{dt}$$

as the expression that force equals rate of change of momentum, analogous to

$$e - L\frac{di}{dt}$$

7. SINE WAVE

The ordinary dynamo is the most familiar application of the principle of induced emf. The field magnets give rise to magnetic flux, and coils of wire (constituting the armature) are caused to move across this flux by some outside source of power.

The simplest type of dynamo generates an alternating current

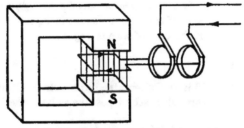

FIG. 6.—*Simple dynamo illustrating how the revolving conductor cuts magnetic lines*

and is sketched in Fig. 6. The single turn of wire shown is in such a position that maximum magnetic flux passes through it. When it is rotated in either direction, the flux passing through it is changed, and hence an electromotive force is induced in it. If the turn of wire is continuously rotated at constant angular speed ω, the rate of change of magnetic flux through it will be greater in some positions than in others, and the electromotive force at the slip-rings AA will vary in a certain manner, this variation being repeated each revolution. In Fig. 7 let POP represent the end view of the turn of wire. As the wire revolves to the successive positions P_1OP_1, P_2OP_2, etc., the emf is proportional to the sine of the angle formed by the revolving wire with POP. The magnitude of the emf at any instant may,

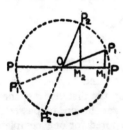

FIG. 7.—*Successive positions of revolving conductor; the emf generated is proportional to the sine of the angle formed by the revolving conductor and POP*

therefore, be represented by the vertical lines P_1M_1, P_2M_2, etc., drawn from the horizontal axis to the end of a line revolving with the angular velocity ω.

A diagram may be drawn, taking the distance along a horizontal line to represent time and the vertical distance from this line to represent induced electromotive force. This emf curve has the mathematical form of a sine wave. Many dynamos in actual use generate electromotive forces very nearly of this form, and on account of its mathematical simplicity the sine wave is assumed in most of alternating-current theory. It should not be forgotten, however, that sine-wave theory is in many practical cases only an approximation because the emf is not rigorously of sine-wave form.

Letting e = emf at any instant, E_0 = maximum emf (that at the crest of the wave as shown in Fig. 8), t = time, ω = angular velocity of the turn of wire in Fig. 6,

$$e = E_0 \sin \omega t \tag{10}$$

This emf alternates in direction. Starting at a, Fig. 8, it passes through a set of positive values, then through a set of negative

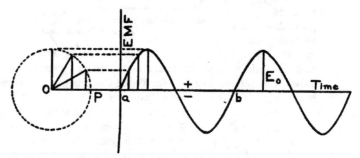

FIG. 8.—*Sine wave developed from circle diagram*

values, and at b begins to repeat the same "cycle." In the time of one complete cycle, represented by the distance $a\,b$, the revolving radius OP makes one complete revolution, or passes through the angle 2π radians. The time required for one complete cycle being represented by T, it follows that

$$\omega = \frac{2\pi}{T}$$

The time T is called the "period" of the alternation. It is the reciprocal of the "frequency," which is the number of times per second that the electromotive force passes through a complete cycle of values. It follows, denoting frequency by f, that

$$\omega = 2\pi f \tag{11}$$

In considering the effect of frequency in electrical phenomena the quantity ω is found more convenient and appears oftener than f.

Effective Values of Alternating Quantities.—The instantaneous rate at which heat is produced in a circuit is proportional to the square of the instantaneous current. According to the equation $p = Ri^2$, the average rate of heat production must be proportional to the average value of i^2. The average heating effect determines the deflection of such an instrument as a hot-wire ammeter, which thus indicates a current I fulfilling the condition, average power $= RI^2$. The indicated current I must therefore be the square root of the average value of i^2. The square root of the mean square of an alternating current or emf is called the "effective" or "root-mean-square" current or emf. All ordinary ammeters and voltmeters used in alternating-current measurements give effective values.

When an electromotive force has the sine-wave form, $e = E_0 \sin \omega t$, the mean square value is proportional to the average value of $\sin^2 \omega t$ during a half cycle, which is equal to 0.5. The effective value is proportional to the square root of this, so that the effective value is $\sqrt{0.5}\, E_0$, or

$$E = 0.707\, E_0$$

Similarly in the case of current,

$$I = 0.707\, I_0$$

8. CIRCUIT HAVING RESISTANCE AND INDUCTANCE

When an emf is suddenly impressed on a circuit containing inductance as well as resistance, say by closing a switch, a current begins to flow but does not rise to its full value instantly. The magnetic flux accompanying the current causes a self-induced emf which by Lenz's law opposes the increase of current. There are then acting in the circuit two emf's, the impressed emf e and the emf of self-induction, which by equation (8) is $-L\dfrac{di}{dt}$. The minus sign is used to indicate that the induced emf opposes the impressed emf.

This is similar to the action of a mechanical force on a material object; the applied force is opposed by the force of inertia and some time is required before the body moves with the final velocity determined by the applied force and the friction. The opposing force of inertia in the mechanical case is given by the product of mass by the time rate of change of velocity. The force of inertia corresponds to induced emf, the mass to inductance, and velocity to current.

The total emf acting to produce current through the resistance is the sum of the impressed emf and the emf of self-induction, thus:

$$e - L\frac{di}{dt} = Ri$$

This equation gives the relation between current and applied emf at any instant. It is usually convenient to consider this equation in the form

$$L\frac{di}{dt} + Ri = e \qquad (12)$$

which indicates that the applied emf is opposed by the resistance and the inductance. When R is relatively large or L relatively small the current comes very quickly to its final value.

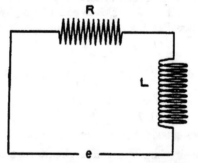

FIG. 9.—*Circuit with resistance and inductance in series*

Impedance.—In alternating-current and radio work the most common and the simplest type of electromotive force is the sine wave. Such an emf is expressed by equation (10). Supposing a sine-wave emf to be impressed on a circuit, equation (12) becomes

$$L\frac{di}{dt} + Ri = E_0 \sin \omega t \qquad (13)$$

The solution of this differential equation (neglecting a term which represents the transient phenomena when the current is started) is

$$i = \frac{E_0}{R^2 + \omega^2 L^2}(R \sin \omega t - \omega L \cos \omega t) \qquad (14)$$

or

$$i = \frac{E_0}{\sqrt{R^2 + \omega^2 L^2}} \sin(\omega t - \theta) \qquad (15)$$

where θ is defined by

$$\tan \theta = \frac{\omega L}{R} \tag{16}$$

The current which flows in a circuit containing constant resistance and inductance due to a sine emf is also a sine wave. The emf varies as sin (ωt), the current as sin $(\omega t - \theta)$, hence the current lags behind the emf by the angle θ. This angle is called the phase angle. The instantaneous current becomes a maximum (I_0) when sin $(\omega t - \theta) = 1$,

$$\therefore I_0 = \frac{E_0}{\sqrt{R^2 + \omega^2 L^2}}$$

and, since the effective values I and E are equal to 0.707 I_0 and 0.707 E_0, respectively,

$$I = \frac{E}{\sqrt{R^2 + \omega^2 L^2}} \tag{17}$$

This has the form of Ohm's law, the quantity $\sqrt{R^2 + \omega^2 L^2}$ occurring in place of R. This quantity is for this circuit the value of the impedance, which is defined as the ratio of emf to current. Since $\omega = 2\pi$ times the frequency it is clear that impedance is a function of frequency as well as of resistance and inductance.

Power.—The power expended in the circuit is at any instant the product of the instantaneous electromotive force and current; $p = ei$. The average power is the mean taken over a complete cycle of this instantaneous product. Performing the calculation, the average power is found to be

$$P = EI \cos \theta,$$

where E and I are effective values and θ is the phase angle, defined by equation (16) above. The ratio $\frac{P}{EI} = \cos \theta$ is called the power factor of the circuit.

9. CIRCUIT HAVING RESISTANCE, INDUCTANCE AND CAPACITY

When a circuit contains a condenser in series with resistance and inductance the applied electromotive force is opposed by the potential difference of the condenser in addition to the opposition of the resistance and inductance. The potential difference of the condenser equals $\frac{Q}{C}$, which may be written $\frac{\int i dt}{C}$, so that equation (12) becomes

$$e = Ri + L\frac{di}{dt} + \frac{\int i dt}{C} \tag{18}$$

Taking $e = E_0 \sin \omega t$, and differentiating the equation,

$$L\frac{d^2 i}{dt^2} + R\frac{di}{dt} + \frac{i}{C} = E_0\,\omega\,\cos\,\omega t$$

The solution (neglecting terms representing the transients, which die out very quickly after the current is started) is

$$i = \frac{E_0}{\sqrt{R^2 + \left(\omega L - \dfrac{1}{\omega C}\right)^2}}\,\sin\,(\omega t - \theta)$$

The phase angle θ is given by

$$\tan\,\theta = \frac{\omega L}{R} - \frac{1}{R\omega C} \tag{19}$$

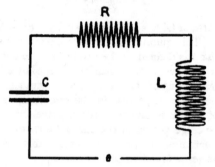

FIG. 10.—*Circuit with resistance, inductance, and capacity in series; a typical radio circuit*

The maximum value of the current is

$$I_0 = \frac{E_0}{\sqrt{R^2 + \left(\omega L - \dfrac{1}{\omega C}\right)^2}}$$

The impedance is

$$\sqrt{R^2 + \left(\omega L - \dfrac{1}{\omega C}\right)^2}.$$

The relation between effective current and emf is

$$I = \frac{E}{\sqrt{R^2 + \left(\omega L - \dfrac{1}{\omega C}\right)^2}} \tag{20}$$

It is to be noted that the terms in L and C have opposite signs in this equation. Thus one tends to neutralize the other, and comparing with equation (17) for a circuit with resistance and inductance only, the impedance of an inductive circuit can be reduced and the current increased by putting a condenser of suitable value in series with the inductance. This increase of the current has sometimes been called resonance or partial resonance, but the term "resonance" is usually reserved for the production of a maximum current, as treated in section 11, below.

Special Cases.—It is of interest to consider the following special cases:[a]

    I. $L = o$ and $C = \infty$.

    II. $C = \infty$.

    III. $L = o$.

Case I represents a circuit with resistance alone. The equations just above give for this case $I = \dfrac{E}{R}$, $\theta = o°$. The impressed electromotive force and the current are in phase, and their ratio is the resistance, just as with direct current.

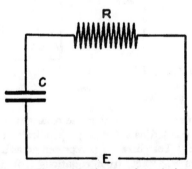

Fig. 11.—*Circuit with resistance and capacity in series*

Case II is that of a circuit with resistance and inductance, which has already been treated in section 8.

Case III is that of a circuit with resistance and capacity in series. Equations (20) and (19) give

$$I = \frac{E}{\sqrt{R^2 + \dfrac{1}{\omega^2 C^2}}}, \quad \tan \theta = -\frac{1}{R\omega C}$$

---

[a] Putting $C = \infty$ is mathematically equivalent to the statement that the condenser is short-circuited. As the distance between the plates of a condenser is decreased the capacity increases without limit. We may consider, then, that when the plates touch together and the condenser is short-circuited the capacity is infinite.

The case is of special importance when the resistance term is entirely due to energy losses within the condenser. It is frequently convenient to deal with the "phase difference" ψ, which $= 90° - \theta$, rather than with the phase angle. The phase difference is given by

$$\tan \psi = - R\omega C$$

If R is very small $\tan \psi = \psi$, and the phase difference $= - R\omega C$.

10. "VECTOR" DIAGRAMS

Writing equation (17) in the form $E = \sqrt{R^2 I^2 + \omega^2 L^2 I^2}$, it is evident that E has such a value as would be given by the diagonal of a rectangle having sides equal to RI and ωLI. It is therefore possible to determine the value of E by the aid of a vector diagram

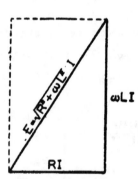
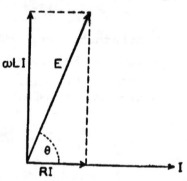

FIG. 12.—*Vector combination of electromotive forces*   FIG. 13.—*Vector diagram for resistance and inductance in series*

such as is used for calculating the resultant of mechanical forces. ωLI is represented as a vector perpendicular to RI, and their resultant is E. The current I is represented as a vector in the same direction as RI; since, from equations (10), (15), and (16), the current and electromotive force differ in phase by the angle whose tangent is $\dfrac{\omega L}{R}$, and this is equal to the angle θ in Fig. 13.

If equation (20) is written in the form

$$E = \sqrt{(RI)^2 + \left(\omega LI - \frac{I}{\omega C}\right)^2},$$

it is evident that E is calculable as the result of adding the three vectors RI, ωLI, and $\dfrac{I}{\omega C}$, drawn as in Fig. 14.

These three quantities are the emf's across the resistance, inductance, and capacity, respectively. The emf $\frac{I}{\omega C}$ is drawn downward from the origin, opposite in direction to ωLI, corresponding to the minus sign in the equation. The phase angle between I and the resultant E is θ. From the Fig. 14,

$$\tan \theta = \frac{\omega L - \frac{I}{\omega C}}{R} \tag{21}$$

in agreement with (19) above. When ωL is greater than $\frac{I}{\omega C}$, E is above the horizontal line as shown, θ is positive, and the cur-

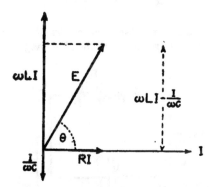

Fig. 14.—*Vector diagram for resistance, inductance, and capacity in series*

rent is said to lag behind the electromotive force. When $\frac{I}{\omega C}$ is greater than ωL, θ is negative, and the current is said to lead the electromotive force. The component of emf in phase with the current is RI. The component at $90°$ to the current is $\left(\omega L - \frac{I}{\omega C}\right) I$. The ratio of this component emf to the current is called the reactance. Its value here is $\left(\omega L - \frac{I}{\omega C}\right)$. The ratio of the resultant emf to the current is called the impedance, which is here equal to $\sqrt{R^2 + \left(\omega L - \frac{I}{\omega C}\right)^2}$.

If all the electromotive forces in Fig. 14 be divided by I, the component vectors then become resistance and reactances and these combine vectorially to give the impedance as a resultant. It is sometimes convenient to speak of resistance and reactance as impedance components. Reactance (usually denoted by the symbol X) is expressible in ohms just as resistance is. The reactance in the case under consideration consists of two parts, the "inductive reactance" ωL and the "capacitive reactance" $\frac{I}{\omega C}$. These may be denoted, respectively, by X_L and X_0. From the expression (21) above it is seen that the tangent of the phase angle is a ratio of impedance components.

$$\tan \theta = \frac{X}{R} = \frac{X_L - X_0}{R}$$

When the capacitive reactance is greater than the inductive reactance, the total reactance X and the phase angle have negative

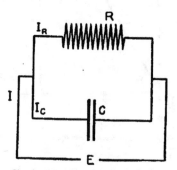

Fɪɢ. 15.—*Circuit with resistance and capacity in parallel*

values. In any case the ratio of reactance to resistance is the tangent of the phase angle.

Phase Difference.—In a circuit consisting of resistance and capacity only, or resistance and inductance only, in series, it is more convenient to deal with the phase difference than with the phase angle. The tangent of the phase difference is the reciprocal of the tangent of the phase angle. When an angle is small it is equal to its tangent, and consequently the phase difference is equal to the ratio of resistance to reactance, in a series circuit in which the resistance is small compared with the reactance. This is in agreement with the case discussed above on page 27, where it was shown that the phase difference of a condenser with resistance in series

$= -R\omega C$. Similarly, the phase difference of an inductance with resistance in series $= \dfrac{R}{\omega L}$.

Vector Addition of Currents.—For series circuits the emfs and the components of impedance combine vectorially just as, with direct current, the emfs and resistances combine algebraically. With direct current in parallel circuits, on the other hand, the currents and conductances add up algebraically. For parallel circuits with alternating current, the currents combine vectorially, and so do the components of the admittance (reciprocal of impedance).

Suppose, for example, a condenser and resistance in parallel (Fig. 15). Current I is the vector sum of the currents $I_R = \dfrac{E}{R}$ and $I_O = \omega CE$. The impressed emf and the current $\dfrac{E}{R}$ are . in the same direction,

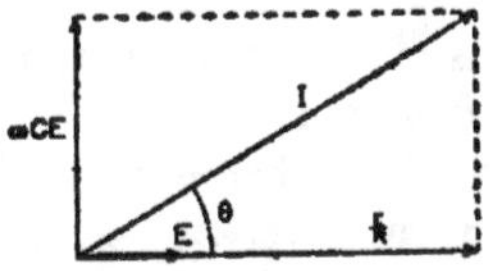

Fig. 16.—*Vector diagram for resistance and capacity in parallel*

while the current ωCE leads the impressed emf by 90°. The resultant current I leads E by the angle θ, where $\tan \theta = R\omega C$.

From Fig. 16,

$$I = E\sqrt{\frac{1}{R^2} + \omega^2 C^2}$$

and the admittance (ratio of resultant current to emf) is $\sqrt{\dfrac{1}{R^2} + \omega^2 C^2}$.

11. RESONANCE

In a circuit consisting of inductance, capacity and resistance in series, the effective current has been shown to be

$$I = \frac{E}{\sqrt{R^2 + \left(\omega L - \dfrac{1}{\omega C}\right)^2}} \tag{22}$$

When $\omega L = \dfrac{1}{\omega C}$, the impedance is a minimum and the effective current is a maximum. This condition for maximum current is called resonance. The ratio of the current at resonance to the current in the circuit with the condenser removed has been called the "resonance ratio"; this quantity is practically the same in radio circuits as the "sharpness of resonance" defined below.

At a given frequency resonance may be brought about by varying either the capacity or the inductance. On the other hand, for a circuit of given L and C, there is some particular frequency at which resonance occurs. The condition

$$\omega L = \frac{1}{\omega C} \tag{23}$$

is equivalent to

$$\left.\begin{aligned} \omega &= \frac{1}{\sqrt{LC}} \\ \text{or, } f &= \frac{1}{2\pi\sqrt{LC}} \end{aligned}\right\} \tag{24}$$

The relation (24) is of the greatest importance in high-frequency work. It is the fundamental equation of the wave meter, for instance. A number of other important ways of expressing the same relation are given in section 78.

Simplified Current Equation at Resonance.—At resonance the inductive reactance is equal to the capacitive reactance, the total reactance is zero, and the impedance equals simply the resistance. That is, at resonance, equation (22) reduces to

$$I_r = \frac{E}{R}$$

This means that the impressed emf is strictly equal to RI_r. The potential difference across the condenser and that across the inductance may be greater than this, and in fact may be many times the impressed emf. Being equal and opposite, they neutralize each other and contribute nothing to the resultant emf opposing the applied emf.

Mechanical Illustration.—The phenomenon of resonance is well

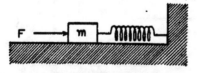

Fig. 17.—*Simple mechanical system which can exhibit the phenomenon of resonance*

illustrated by the vibration of a spring with a mass attached. When a force F acts on the mass m, it is opposed by the stiffness of the spring, by the inertia of the mass, and by friction. The analogy to the electrical case is not perfect, since friction due to sliding is not proportional to the velocity. If the force is applied periodically, there will be a certain particular frequency for which a more vigorous oscillation is produced than for any other. When the frequency of the applied force is just equal to the frequency of resonance, the applied force is all used in overcoming friction; the elasticity of the spring and the inertia of the mass constitute two equal forces opposing each other. These two opposite forces may be much greater than the applied force. For instance, the vibration may become so violent as to break the spring although the impressed force is far too small to do so.

Magnification of Voltage.—Similarly, in the case of the electrical circuit, there is danger of breaking down the condenser in a resonant circuit because the potential difference across the condenser is much greater than the applied electromotive force. The ratio of the voltage across the condenser to the applied voltage is greater the smaller the resistance (including under resistance not only the ordinary "ohmic" resistance of the conductors but all sources of energy loss, such as dielectric loss in the condenser). In comparing the electrical circuit with the vibrating spring, one should remember that the mass is the analog of the inductance and the spring the analog of the condenser. It is unfortunate that the diagram generally used for an inductance is the same as that used above for a spring.

Vector Diagram of Resonance.—Resonance phenomena are shown in an interesting manner by means of vector diagrams. In Fig. 14, which illustrates the vector diagram of emfs for a circuit with resistance, inductance, and capacity in series, the inductive reactance ωL is taken to be greater numerically than the capacitive reactance $\frac{1}{\omega C}$ so that the resultant vector E has a direction corresponding to a positive rotation from the direction of RI through an angle θ. Suppose that the frequency is decreased; ωL decreases and $\frac{1}{\omega C}$ increases numerically. When they become equal, the total reactance $\left(\omega L - \frac{1}{\omega C}\right)$ and the angle θ become zero, and E equals RI. The diagram then becomes Fig. 18.

Resonance Curves.—As already stated, in a circuit of given L and C there is some frequency at which resonance occurs, given by $\omega L = \frac{1}{\omega C}$. At all other frequencies the inductive reactance and the capacitive reactance are unequal, and their difference enters the expression for impedance. At frequencies less than the frequency of resonance the capacitive reactance is the larger, and consequently we may say that at low frequencies the capacity keeps down the current, while at high frequencies it is the inductance that keeps the value of the current down. For any departure from the condition of resonance, whether by variation of frequency, of inductance, or of capacity, the current is diminished. The process of varying either the capacity or the inductance to obtain the setting at which the circuit is in resonance with the frequency of the applied electromotive force is called "tuning" the circuit or tuning the circuit to resonance.

The reduction of the current on both sides of resonance is shown in Fig. 19, in which the square of current is plotted against capacity, the emf being constant. Such curves are called resonance curves.

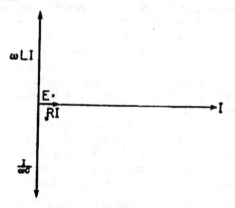

FIG. 18.—*Vector diagram of series circuit in resonance*

The three curves shown are for an actual circuit, with its normal resistance of 4.4 ohms, with 5 ohms added, and with 10 ohms added. The inductance is 377 microhenries and the frequency 169 100 cycles per second. The curves show theoretical values as obtained from the formula

$$I^2 = \frac{E^2}{R^2 + \left(\omega L - \dfrac{1}{\omega C}\right)^2}$$

The theoretical values were closely checked by actual observations, using a pliotron as a source of alternating current and a thermocouple and galvanometer to measure the current. The square of current is plotted instead of current; simply because the galvanometer deflections were proportional to the square of current. The ordinates are thus in terms of galvanometer deflections.

The value of the constant impressed emf is given by $E = RI_r$, where $I_r =$ current at resonance. In the arbitrary units resulting from the expression of current-square in galvanometer deflections and R in ohms, for curve A the value of $E = 4.4 \times \sqrt{19} = 19.2$. The emf across the inductance $= \omega LI$, so that at resonance its

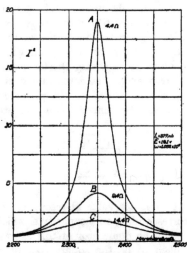

FIG. 19.—*Resonance curves for series circuit with different resistances*

value is $2\pi \times 169$ $100 \times 377 \times 10^{-6} \times \sqrt{19} = 1750$. The emf across the condenser at resonance $= \dfrac{I_r}{\omega C_r}$, which equals the same value, 1750. Note that this is much greater than the applied emf, 19.2. In the case of curve B, the applied emf is the same, but the emf across the inductance and the equal emf across the condenser at resonance $= 2\pi \times 169$ $100 \times 377 \times 10^{-6} \times \sqrt{4.16} = 818$. In the case of curve C, the equal emfs across inductance and capacity at resonance each equal 534. The applied emf having the same value, 19.2, for each curve, this clearly illustrates the statement previously made that the ratio of condenser voltage to applied voltage is greater the smaller the resistance.

Sharpness of Resonance.—One of the principal applications of the phenomenon of resonance is the determination of frequency. Since the current is a maximum when $\omega = \dfrac{1}{\sqrt{LC}}$, the frequency is determined when L and C are known. The precision with which frequency can be determined by this method depends upon the sensitiveness of the current indication to a given change in C or L at resonance. This sensitiveness is obviously greater the sharper the peak (Fig. 19). The precision of determination of frequency, therefore, depends upon what may be called the sharpness of resonance, a quantity which measures the fractional change in current for a given fractional change in either C or L at resonance. (This quantity has also been called selectivity; see also statement on p. 31 regarding the term resonance ratio.)

The sharpness of resonance is an important characteristic of a circuit and is very simply related to the phase differences and other constants. It may be defined in mathematical terms by the following ratio

$$\frac{\sqrt{\dfrac{I_r^2 - I_1^2}{I_1^2}}}{\dfrac{\pm (C_r - C)}{C}}$$

where the subscript r denotes value at resonance, and I_1 is some value of current corresponding to a capacity C which differs from the resonance value. The numerator of this expression is somewhat arbitrarily taken to be the square root of the fractional change in the current-square instead of taking directly the fractional change of the first power of current. This is done because of the convenience in actual use of this expression (since the deflections of the usual detecting devices are proportional to the square of the current) and also because of its mathematical convenience. It is readily shown that the sharpness of resonance thus defined is equal to the ratio of the inductive reactance to the resistance. Since

$$\omega L = \frac{1}{\omega C_r} \qquad (25)$$

the relation

$$I_1^2 = \frac{E^2}{R^2 + \left(\omega L - \dfrac{1}{\omega C}\right)^2}$$

becomes

$$I_1^2 = \frac{E^2}{R^2 + \left(\dfrac{1}{\omega C_r} - \dfrac{1}{\omega C}\right)^2}$$

This, together with

$$I_r^2 = \frac{E^2}{R^2}$$

gives

$$\frac{I_r^2}{I_1^2} = \frac{R^2 + \left(\dfrac{1}{\omega C_r} - \dfrac{1}{\omega C}\right)^2}{R^2}$$

$$\frac{I_r^2 - I_1^2}{I_1^2} = \frac{\dfrac{1}{\omega^2}\left(\dfrac{1}{C_r} - \dfrac{1}{C}\right)^2}{R^2}$$

$$= \frac{1}{R^2\omega^2}\left(\frac{C - C_r}{C_r\,C}\right)^2$$

$$\frac{\sqrt{\dfrac{I_r^2 - I_1^2}{I_1^2}}}{\dfrac{\pm (C_r - C)}{C}} = \frac{1}{R\omega C_r} \tag{26}$$

The right-hand member of the equation is the ratio of the capacitive reactance at resonance to the resistance. In virtue of equation (25), it is also equal to the ratio of the inductive reactance to the resistance; thus sharpness of resonance $= \dfrac{\omega L}{R}$.

It is of interest to note the relation of the sharpness of resonance to phase difference. As shown above on page 31, the phase difference of a series combination either of resistance and inductance or of resistance and capacity is equal to the ratio of resistance to reactance. If in a circuit having an inductance coil and a condenser in series the only resistance is that of the inductance coil, it follows that the sharpness of resonance is equal to the reciprocal of the phase difference of the coil. If, on the other hand, the resistance of the circuit is all due to energy loss in the condenser, the sharpness of resonance is equal to the reciprocal of the phase difference of the condenser. A measurement of the sharpness of resonance thus gives the phase difference directly. If the resistance is partly in the coil and partly in the condenser, each has a phase difference and the sharpness of resonance is equal to the reciprocal of the sum of the two phase differences.

It has been mentioned that there is danger of breaking down a condenser in a resonant circuit because the potential difference across the condenser may be many times higher than the applied electromotive force. This danger is directly in proportion to the sharpness of resonance, or inversely as the phase difference of the condenser. For if the applied emf is RI, the condenser voltage at resonance is $-\dfrac{I}{\omega C_r}$. The ratio of the condenser voltage to the applied emf is $-\dfrac{1}{R\omega C_r}$, which is the sharpness of resonance or the reciprocal of the condenser phase difference.

Application to Radio Resistance Measurement.—Formula (26) above, which gives the relation between the sharpness of resonance

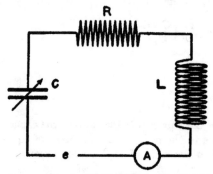

R

C

L

e   A

FIG. 20.—*Simple circuit for measurements of resistance or wave length*

and the phase difference of the condenser, has been shown to have important applications to the precision of frequency measurement and to the rise of voltage on the condenser. Another application of great importance is the measurement of resistance. This is seen by writing the equation in the form

$$R = \frac{\pm(C_r-C)}{\omega C_r C}\sqrt{\frac{I_1^{\,2}}{I_r^{\,2}-I_1^{\,2}}} \qquad (27)$$

Thus the resistance of a simple circuit as in Fig. 20 is measured by observing deflections of the indicating instrument A for two settings of the variable condenser, one setting at resonance C_r and any other setting C. This is one of the principal methods of measuring high-frequency resistance, and may be called the "reactance variation" method. Other ways of using the principle of reactance variation are described in section 50 below.

The method is rigorous, involving no approximations, provided the applied emf is undamped. The resistance so measured is the effective resistance of the entire circuit, including that due to condenser losses and radiation.

12. PARALLEL RESONANCE

When a coil and a condenser are in parallel in a circuit, the phenomena are strikingly different from those of the series arrangement. The total current I is the vector sum of the currents in the two branches, I_L and I_C. The current through the coil depends on its resistance and inductance, thus,

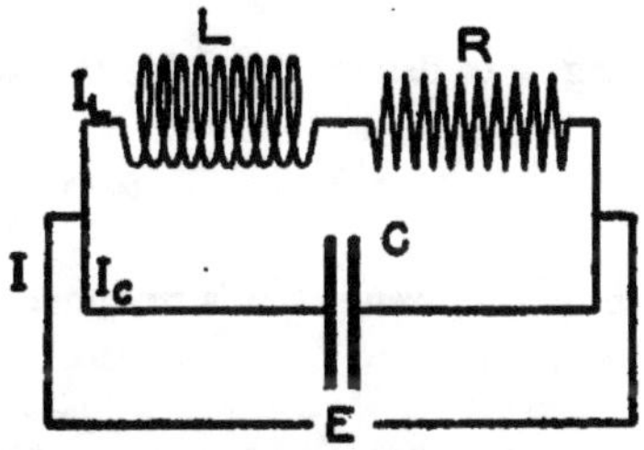

$$I_L = \frac{E}{\sqrt{R^2 + \omega^2 L^2}}$$

Also

$$I_0 = -\omega C E$$

FIG. 21.—*Parallel circuit having capacity in parallel with inductance and resistance; under certain conditions the current in either of the branches exceeds that in the main line*

assuming the condenser loss to be negligible. Taking the sum of the two currents, with due regard to their phase relation, the total current is

$$I = E \sqrt{\left(\omega C - \frac{\omega L}{R^2 + \omega^2 L^2}\right)^2 + \left(\frac{R}{R^2 + \omega^2 L^2}\right)^2}$$

When

$$\omega C = \frac{\omega L}{R^2 + \omega^2 L^2} \tag{28}$$

the total current is in phase with the emf, and has the value,

$$I_r = \frac{ER}{R^2 + \omega^2 L^2} \tag{29}$$

This is the minimum current for varying values of C and is very nearly the minimum current for varying values of L or ω.

Equation (28) is the condition for what may be called inverse resonance or parallel resonance. At parallel resonance, the total current in the external circuit is less than the current in the coil. This is because the currents in condenser and coil are in

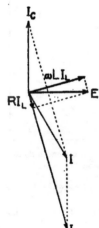

FIG. 22.—*Vector diagram for parallel circuit*

opposite directions as regards the external circuit, and thus tend to neutralize each other in that circuit.

Simple Case.—When the resistance of the coil is very small compared with its reactance, as is usual at radio frequencies, equation (28), the condition for parallel resonance, becomes

$$\omega_o C = \frac{1}{\omega_o L}$$

(using ω_o to denote the value of ω at parallel resonance), or

$$\omega_o = \frac{1}{\sqrt{LC}} \tag{30}$$

The total current at this frequency is

$$I_r = \frac{ER}{\omega_o^2 L^2} \tag{31}$$

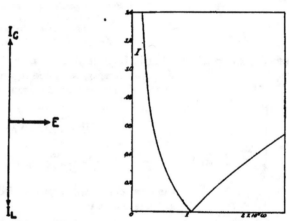

FIG. 23.—*Vector diagram illustrating resonance in simple parallel circuit*

FIG. 24.—*Resonance curve showing the condition of parallel resonance*

The current in the condenser is very closely equal to that in the coil, the value being $\frac{E}{\omega_o L}$. The total current is the vector sum of the currents in coil and condenser, and is thus smaller than the current in either by the ratio $\frac{R}{\omega_o L}$. As suggested by a comparison of Fig. 22 and Fig. 23 the resultant current may be vanishingly small. The combination of coil and condenser acts like a very large impedance in the main circuit, the value of this impedance being $\frac{\omega_o{}^2 L^2}{R}$. For any variation of frequency, inductance, or capacity, from the condition of parallel resonance, the total current increases.

At low frequencies the inductance carries the greater part of the current, and at high frequencies the condenser is the more important.

Comparison of Series and Parallel Resonance.—It is interesting to compare the resonance phenomena in a series circuit with the phenomena of parallel resonance. In the former case the individual voltages across the coil and condenser exceed the resultant voltage across both, whereas in the latter case the separate currents exceed the resultant current. The impedance introduced into the circuit by the series combination is vanishingly small, and the impedance due to the parallel combination is very large. Comparing equation (29) with $I = \dfrac{E}{\sqrt{R^2 + \omega^2 L^2}}$, the ratio of the total current at parallel resonance to the current in the circuit with the condenser removed is $\dfrac{R}{\sqrt{R^2 + \omega^2 L^2}}$. Thus the current is reduced in parallel resonance in the same ratio that it is increased in series resonance. The two kinds of resonance are discussed further in the next section.

RADIO CIRCUITS

13. SIMPLE CIRCUITS

A typical radio circuit comprises an inductance coil, a condenser, and a source of electromotive force, in series.

The source E may be a small coil in which an alternating electromotive force is induced by the current in a neighboring circuit. Some of the phenomena in such a circuit have already been treated under "The principles of alternating currents." What is there given applies to high as well as low frequencies. Some of the phenomena and their mathematical treatment are much simplified at high frequencies. Electromotive forces of sine-wave form are assumed in this discussion; the results obtained apply equally to slightly damped waves.

It will be recalled that the reactance of an inductance is ωL, and the reactance of a capacity is $\dfrac{1}{\omega C}$. It is essential to remember that ω is 2π times the frequency. In fact, the physical meaning of ω in reactance expressions is the same thing as frequency; the 2π is a factor resulting from the way the units are defined. The expression ωL tells us that the reactance of an inductance is proportional to frequency.

Series Circuit.—The simple circuit of Fig. 25 is, in fact, the principal circuit used in radio transmitting sets, receiving sets, and wave meters. Some of its chief properties are conveniently brought out by a graphical study of the variation of its reactance with frequency. Advantage is taken of the fact that resistance is a negligible part of the impedance (except at resonance), to obtain very simply an idea of the way the current varies with frequency. Small current corresponds to large reactance (either positive or negative), and vice versa.

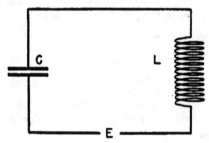

Fig. 25.—*Simple series circuit*

The reactance of the circuit is $\left(\omega L - \dfrac{1}{\omega C}\right)$. The inductive reactance, ωL, is the predominating portion of this at high frequencies. It is represented by the line ωL in Fig. 26. The capacitive reactance predominates at low frequencies; it is represented by the line $\dfrac{1}{\omega C}$. The sum of these two is represented by the line marked total reactance. At the point ω' where this reactance curve crosses the axis—i. e., where $\omega L = \dfrac{1}{\omega C}$—the current is a maximum.

The resonance curve, showing variation of the current with frequency, would rise to infinity at the point ω', where the total reactance is o, if it were not for the resistance in the circuit. While the current at resonance is determined by the resistance, the frequency of resonance is given accurately by the reactance curve which takes no account of resistance. The most important

aspect of resonance phenomena is thus shown by the simple reactance curve, the plotting of the current curve being unnecessary.

Use of Reactance Curves.—Complex circuits can be studied and
much useful information easily obtained by the use of reactance
curves. The effect of any auxiliary circuit upon a wave meter or
a transmitting apparatus can be determined, as will be shown
later. In any such diagram the points where the reactance curve
crosses the ω axis give the frequencies at which the current is a
maximum.

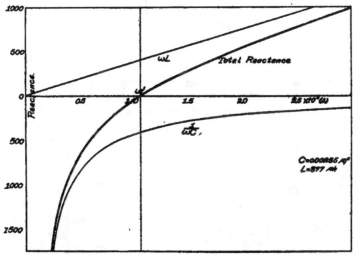

Fig. 26.—*Reactance diagram for simple series circuit, showing the capacitive and inductive reactances and their resultant at different frequencies. The current in the circuit
is a maximum at ω'*

Parallel Circuit.—An inductance and a capacity placed in parallel in a circuit behave very differently from the series arrangement
of inductance and capacity already discussed. As shown in Fig.
27, the same electromotive force is impressed upon the terminals
of both L and C by the source E, which may be a spark gap,
another condenser, a coupling coil, etc. The total current I is the
sum of the currents in L and C, or

$$I = \frac{E}{\omega L} - \omega C E$$

The ratio, $\frac{I}{E}$, is equal to the reciprocal of the impedance, and when resistance is negligible this is a quantity called the susceptance. The total susceptance is here made up of two parts, the inductive susceptance $\frac{1}{\omega L}$, which predominates at low frequencies, and the

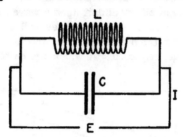

Fig. 27.—*Simple parallel circuit*

capacitive susceptance ωC, which predominates at high frequencies. Each of these two susceptances is the reciprocal of the corresponding reactance, but this is true only when resistances are negligible.

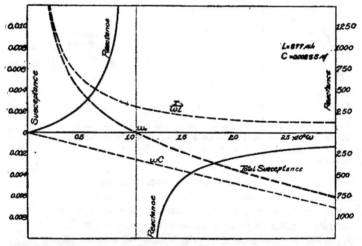

Fig. 28.—*Reactance diagram for simple parallel circuit, showing the capacitive and inductive susceptances at different frequencies, together with their resultant and the resultant reactance*

The curve marked "Total susceptance" in Fig. 28 was obtained by addition of the two curves, $\frac{1}{\omega L}$ and ωC. The curve "Reactance" was obtained by taking reciprocals of the points on the curve of total susceptance. The reactance of the circuit is small at very low and very high frequencies, but at ω_0, the point of parallel resonance, both branches of the reactance curve go to infinity. The current in the circuit is a minimum at ω_0 and would be strictly zero if there were actually no resistance in either branch of the circuit.

Thus it is seen that while a series combination of inductance and capacity has zero reactance when $\omega L = \dfrac{I}{\omega C}$, a parallel arrangement has infinite reactance under the same condition. A series arrangement is therefore used when it is desired to make current of a given frequency a maximum, and a parallel arrangement is used when it is desired to suppress the current of that frequency.

14. COUPLED CIRCUITS

Circuits which are more complex than those already discussed may be considered as combinations of simple circuits. The component simple circuits in general have certain parts in com-

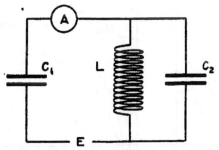

FIG. 29.—*Simple case of coupled circuits in which a parallel circuit is combined with a series circuit*

mon, and these parts are said to constitute the coupling between the circuits. Suppose, for example, the simple series circuit and the simple parallel circuit are combined as shown in Fig. 29. The coil L is the coupling between the circuit C_1L and the circuit C_2L.

Elimination of Interference.—A great deal of information about coupled circuits may be obtained from their reactance diagrams. A curve of the variation of reactance with frequency tells in a very simple way at what frequencies the current is either large or small. The reactance of C_2 and L in parallel (Fig. 29) is as shown in Fig. 28 and designated by X'' in Fig. 30. This combination is in series with C_1. Adding the curve of condenser reactance $\frac{1}{\omega C_1}$ to the curve $X_\bullet''$, the curve X is obtained, giving the reactance to current flowing through the ammeter. At the frequency corresponding to ω', the reactance is zero and the current a maximum. At ω_0 the reactance is infinite and the current is a minimum. It is easily seen, therefore, that such a circuit is very useful where it is desired to have current of a certain frequency in a circuit but to exclude current of a certain other frequency. For example, if it is desired to receive radio messages of a certain wave length from a distant station, and a

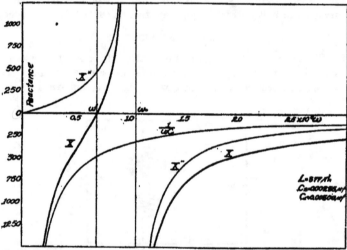

Fig. 30.—*Reactance diagram for combination circuit of Fig. 29; curve X is the resultant reactance of the system*

near-by station operating on a different wave length emits waves so powerful as to interfere with the reception, the interfering signals can be greatly reduced by using this kind of circuit. The circuit C_2L is first independently tuned to resonance with the waves which it is desired to suppress. The setting of condenser C_1 is then varied until the main circuit is in resonance with the desired waves. If the resistances in the circuit are very small, interference is readily eliminated in this manner. The same thing is accomplished by other types of coupled circuits, as explained below.

Suppression of Harmonics.—Such a circuit is useful also in sending stations or in laboratory set-ups, where certain wave lengths need to be eliminated. For example, the emf from an arc generator is not a pure sine wave but contains harmonics in addition to the fundamental frequency. Some harmonic may be especially strong and it may be desired to suppress it. This can be accomplished in some cases by connecting a condenser across a loading coil (which is not a coil used to introduce the emf into the circuit) either in the closed circuit or the antenna and tuning the combination of loading coil and condenser to the objectionable frequency. Various modifications of this simple scheme can be used, which may be more convenient under certain circumstances. Thus, instead of a condenser only, a condenser and coil in series can be connected around the main inductance as in Fig. 31. The circuit L_bMC_2 is independently tuned to

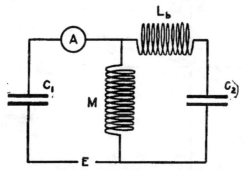

Fig. 31.—*Coupled circuits involving two simple series circuits; various modifications of this circuit are used in transmitting sets in which it is desired to emit certain frequencies and suppress others*

the harmonic which is to be suppressed. The main circuit is then tuned to the frequency which is to be emitted. The reactance to the emitted frequency is thus made zero, while the reactance to the objectionable frequency is made very large, as shown in Fig. 32. The reactance of the parallel combination of M with L_b and C_2 is found by the method used before to be the curve X'' with two branches. The condenser reactance $\dfrac{1}{\omega C_1}$ is added to this, giving the heavy curve X of reactance to current flowing through the ammeter.

One of the characteristic properties of coupled circuits is brought out by Fig. 32, viz, the reactance is zero at two frequencies. That is the current is a maximum for two different frequencies. Between these two, indicated by ω' and ω'', is the frequency of infinite reactance or minimum current, indicated by ω_0. Thus, it is possible to suppress a certain frequency and tune the circuit to a different frequency either larger or smaller than the one suppressed. There will be current maxima at *both* frequencies (corresponding to ω' and ω'') in the circuit if the source of emf supplies these two frequencies simultaneously. The current will have a minimum at the intermediate frequency of infinite reactance only provided the resistances of the circuits are small.

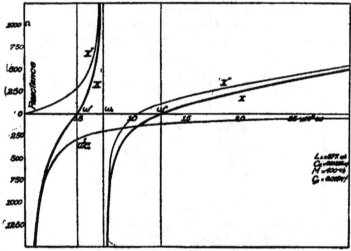

FIG. 32.—*Reactance diagram for the simple coupled circuits of Fig. 31; the curve X of resultant reactance is zero for two values of frequency*

15. KINDS OF COUPLING

Circuits may be connected or coupled together in a number of ways. When there are two circuits, the one containing the source of power is called the primary, the other circuit the secondary. These are generally coupled in one of the following ways: (*a*) By direct connection across an inductance coil; (*b*) by electro-magnetic induction; (*c*) by direct connection across a condenser. In the first kind, called "direct coupling," an inductance coil is common to the two circuits as illustrated in Fig. 33 (*a*). In the second kind, "inductive coupling," shown in (*b*), the two circuits are connected only by mutual inductance. An example of the third kind, "capacitive coupling," is shown in (c); a condenser is common to both circuits in place of the coil M of Fig. 33 (a).

It is characteristic of coupled circuits that the impedances in each circuit affect the current flowing in the other. This reaction of the circuits upon each other is the more marked when the common portion of the two circuits is a larger proportion of their impedances. When this is large the coupling is said to be "close" and when small the coupling is "loose." In the case of extremely loose coupling, the back action of the secondary on the primary circuit is negligible, and the considerations of coupled circuits do not apply; the two circuits act practically as independent circuits, the primary merely applying an electromotive force to the secondary.

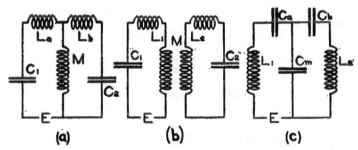

Fig. 33.—*Types of coupling; (a) direct coupling, (b) inductive coupling, (c) capacitive coupling*

Coupling Coefficient.—The closeness of coupling is specified by a quantity called the coupling coefficient. This is defined as the ratio $\dfrac{X_m}{\sqrt{X_1 X_2}}$, where X_m is the mutual or common reactance (either inductive or capacitive) and X_1 is the total inductive or capacitive reactance in the primary circuit and X_2 the total similar reactance in the secondary. Thus, in the case of direct coupling, Fig. 33 (a), the coupling coefficient is

$$\frac{\omega M}{\sqrt{\omega(L_a + M)\omega(L_b + M)}}.$$

Denote the total inductance of primary and secondary by L_1 and L_2, respectively, and the coupling coefficient by k; then

$$k = \frac{M}{\sqrt{L_1 L_2}}$$

This also gives the coupling coefficient for inductively coupled circuits, as illustrated in Fig. 33 (b), L_1 and L_2 being the respective total inductances of primary and secondary, each measured with the other circuit removed. As suggested by the identity of expression for coupling coefficient, inductively coupled circuits may be considered as equivalent to direct-coupled circuits having the same M, C_1, and C_2, and in which $L_a = L_1 - M$ and $L_b = L_2 - M$.

The coupling coefficient in Fig. 33 (c) is:

$$\frac{\dfrac{1}{\omega C_m}}{\sqrt{\left(\dfrac{1}{\omega C_a}+\dfrac{1}{\omega C_m}\right)\left(\dfrac{1}{\omega C_b}+\dfrac{1}{\omega C_m}\right)}}$$

Denote by C_1 the total capacity of the primary circuit, and by C_2 the total capacity in the secondary.

$$\frac{1}{C_1}=\frac{1}{C_a}+\frac{1}{C_m} \text{ and } \frac{1}{C_2}=\frac{1}{C_b}+\frac{1}{C_m}$$

whence,

$$k=\frac{\sqrt{C_1 C_2}}{C_m}.$$

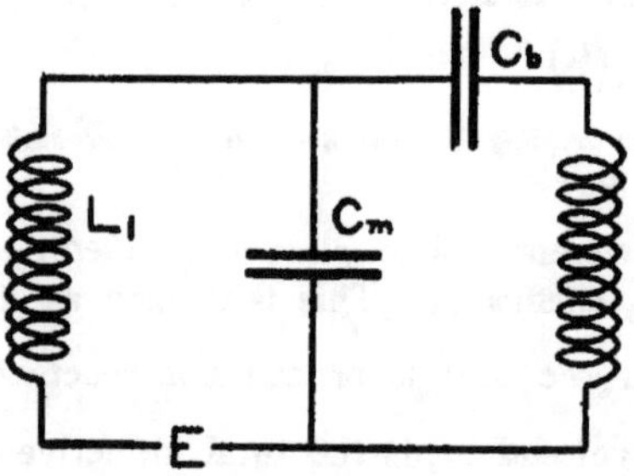

FIG. 34.—*Special case of capacitive coupling*

From these expressions the coupling coefficient may be obtained for particular cases. Thus, for Fig. 31, which is a special case of the kind of coupling shown in Fig. 33 (a),

$$k=\sqrt{\frac{M}{L_b+M}}.$$

Similarly, Fig. 34 shows a special case of capacitive coupling.

The coupling coefficient is readily found to be

$$k = \sqrt{\frac{C_b}{C_b + C_m}}.$$

Use of Coupled Circuits to Select Frequencies.—Any of the systems of coupled circuits which have been mentioned may be used for the purpose of suppressing current of one frequency while responding to current of another frequency or wave length. This was discussed above in connection with the simple case of direct coupling in Fig. 29. It may be shown that each of the more general circuits in Fig. 33 will accomplish the same thing. This

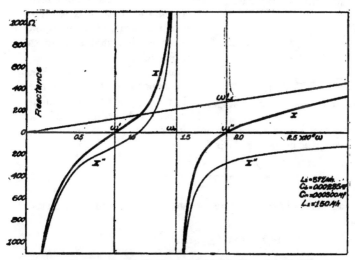

FIG. 35.—*Reactance diagram for case of capacitive coupling shown in Fig. 34*

is also true of the simple case of capacitive coupling in Fig. 34, as may be seen from its reactance diagram Fig. 35. The curve X'' gives the reactance of the parallel combination of C_m with L_2 and C_b. Adding the reactance of L_1 to this, the heavy curve X is obtained, showing the total reactance to current in the primary circuit. As before, the reactance is zero at two frequencies and is infinite at one intermediate frequency.

Thus any of these arrangements of coupled circuits may be used to remove an objectionable frequency while tuning to some other frequency either higher or lower than the one suppressed, provided the resistances of the circuits are not large. In every case, ω_0 corresponding to the frequency suppressed in the primary circuit is given by

$$\omega_0 = \frac{I}{\sqrt{L_2 C_2}}$$

where L_2 =total inductance of the secondary circuit and C_2 = total capacity.

16. DIRECT COUPLING

The above discussion shows how a qualitative comprehension of the action of coupled circuits may readily be obtained. The exact frequencies to which a coupled system responds may be obtained by calculation in the manner here shown for direct coupling, upon the assumption that resistances can be neglected.

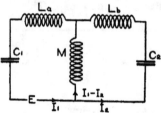

FIG. 36.—*Circuits involving direct coupling*

The emf E in the primary (Fig. 36) is opposed by the impedance of L_a, of C_1, and of the parallel combination of M with L_b and C_2.

Denoting by I_1 and I_2 the currents in primary and secondary, respectively, the current in M is $I_1 - I_2$.

$$E = \left(\omega L_a - \frac{I}{\omega C_1} \right) I_1 + \omega M (I_1 - I_2) \tag{32}$$

The emf across M is the same as that across L_b and C_2 in series, hence

$$\omega M (I_1 - I_2) = \left(\omega L_b - \frac{I}{\omega C_2} \right) I_2$$

$$= \frac{\omega M \left(\omega L_b - \frac{I}{\omega C_2} \right)}{\omega (L_b + M) - \frac{I}{\omega C_2}} I_1$$

Therefore,

$$\frac{E}{I_1} = \omega L_a - \frac{I}{\omega C_1} + \frac{\omega M \left(\omega L_b - \frac{I}{\omega C_2} \right)}{\omega (L_b + M) - \frac{I}{\omega C_2}} \tag{33}$$

The last term is X'', the reactance of the parallel combination of M with L_b and C_2, which may be shown, as before, to vary with frequency according to the curve marked X'' in Fig. 32. Adding to this the curve of $\omega L_a - \dfrac{1}{\omega C_1}$, the total reactance curve X is obtained (Fig. 37). This curve is the graph of equation (33). The value of ω at which $\dfrac{E}{I_1}$ is ∞, or the current in the primary a minimum, is obtained when the denominator of the last term is zero,

$\omega(L_b + M) - \dfrac{1}{\omega C_2} = 0$. Thus $\omega_2 = \dfrac{1}{\sqrt{(L_b + M)C_2}}$. The symbol ω_2 is used to indicate that this is the value for resonance in the secondary circuit $C_2 L_b M$ when the primary circuit is open.

The values of ω at which the primary current is a maximum are given by equating (33) to 0 and solving for ω. A similar expression involving the secondary current may be treated in the same way, and it is found that the secondary current has maxima at the same values of ω as the primary current. Expressed in terms of the inductances and capacities, the solution is rather complicated. It is more convenient to express it in terms of k, the coupling coefficient, and ω_1 and ω_2, the respective values of ω for resonance in the primary circuit $C_1 L_a M$ alone and in the secondary circuit $C_2 L_b M$ alone.

Using the relations,

$$k = \dfrac{M}{\sqrt{(L_a + M)(L_b + M)}}, \ \omega_1 = \dfrac{1}{\sqrt{(L_a + M)C_1}}, \ \omega_2 = \dfrac{1}{\sqrt{(L_b + M)C_2}}, \quad (34)$$

the following two values are found for which the currents have maxima,

$$\omega' = \sqrt{\dfrac{\omega_1^2 + \omega_2^2 - \sqrt{(\omega_1^2 - \omega_2^2)^2 + 4k^2\omega_1^2\omega_2^2}}{2(1 - k^2)}} \quad (35)$$

$$\omega'' = \sqrt{\dfrac{\omega_1^2 + \omega_2^2 + \sqrt{(\omega_1^2 - \omega_2^2)^2 + 4k^2\omega_1^2\omega_2^2}}{2(1 - k^2)}} \quad (36)$$

Example.—The theory was experimentally verified in the following case. Two circuits were direct-coupled as in Fig. 36. The following capacities and inductances were used:

$$C_1 = 0.0023 \times 10^{-6} \text{ farad}$$
$$C_2 = 0.00093 \times 10^{-6} \text{ farad}$$
$$L_a = 56 \times 10^{-6} \text{ henry}$$
$$L_b = 209 \times 10^{-6} \text{ henry}$$
$$M = 241 \times 10^{-6} \text{ henry}$$

The coupling coefficient and the values of ω for resonance in the primary circuit alone and in the secondary circuit alone are found by (34) to be

$$k = 0.659$$
$$\omega_1 = 1.210 \times 10^6$$
$$\omega_2 = 1.545 \times 10^6$$

The frequencies for maximum current in the coupled system are found by (35) and (36) to be

$$\omega' = 1.037 \times 10^6$$
$$\omega'' = 2.392 \times 10^6$$

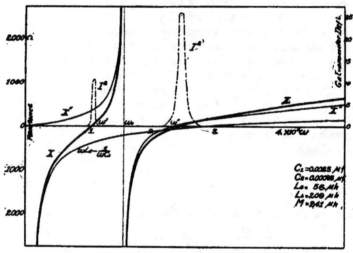

FIG. 37.—*Reactances and current in case of direct coupling*

It is thus evident that the effect of coupling is to spread out or separate farther the two independent frequencies. Values of primary reactance were calculated by equation (33) for a number of values of ω, giving the curve X of Fig. 37. It will be noted that it crosses the ω axis at the values just given for ω' and ω'', and that it goes to $+$ and $-$ infinity at 1.545×10^6, the value of ω_2.

Current was produced in the primary circuit by induction from a buzzer, the buzzer circuit being varied to supply different frequencies. The coil L_a was inductively coupled to the buzzer circuit, the coupling with that circuit being so loose that the emf could be considered as applied at one point of the circuit $L_a MC_1$.

Current was measured by a galvanometer and a crystal detector attached to a circuit inductively coupled to the secondary circuit $L_b\ MC_2$. The galvanometer deflections were approximately proportional to the square of the current. As shown, the curve of observed galvanometer deflections for varying frequency has two maxima corresponding closely to ω' and ω''. The slight discrepancies are probably due to inaccuracies in the values used for L_1 and L_2; the inductances of these coils were later found to vary slightly with frequency, whereas a constant value was assumed for each in computing the reactance curve.

Special Cases.—In the special case when $\omega_1 = \omega_2$, equations (35) and (36) become

$$\omega' = \frac{\omega_1}{\sqrt{1+k}} \tag{37}$$

$$\omega'' = \frac{\omega_1}{\sqrt{1-k}} \tag{38}$$

When k is very small; that is, when M is very small compared with L_a or L_b,

$$\omega' = \omega'' = \omega_1 \tag{39}$$

In this case, where the coupling is very loose, the system responds to only one frequency instead of two, and this is the frequency of resonance of either circuit by itself.

When, on the other hand, L_a and L_b are very small compared with M, the coupling is said to be very close, and k approaches the value unity. As k increases to this value, the two frequencies become more widely separated and in the limit

$$\omega' = \frac{1}{\sqrt{2}}\,\omega_1 \tag{40}$$

$$\omega'' = \infty \tag{41}$$

Practically this means that when $\omega_1 = \omega_2$ and L_a and L_b are negligible in comparison with M, there is only one frequency and this is given by $\omega' = \frac{1}{\sqrt{2MC_1}}$. The reactance curve of such a system is of the type shown in Fig. 30. The curve X crosses the ω axis at ω', a value less than ω_1 (called ω_0 in the figure), and touches the ω axis again at infinity.

A particularly interesting special case of direct coupling is that in which $\omega_r = \omega_2$ and $L_1 = L_2$. This is obtained when $L_a = L_b$ and

$C_1 = C_2$. The values of ω for maximum current in the primary are

$$\omega' = \frac{1}{\sqrt{(L_a + 2M)C_1}} \tag{42}$$

$$\omega'' = \frac{1}{\sqrt{L_a C_1}} \tag{43}$$

In this case one of the frequencies is constant, not varying when L_a is kept constant and M is varied. In the reactance diagram, Fig. 37, the point ω'' remains fixed, and ω' moves farther to the left as M is increased. When M is extremely small, the two frequencies are equal, and the equations (42) and (43) reduce to (39). When M is very large in comparison with L_a, the equations reduce to (40) and (41).

17. INDUCTIVE COUPLING

The applied emf E in the primary of two inductively coupled circuits must satisfy:

$$E = \left(\omega L_1 - \frac{1}{\omega C_1}\right) I_1 - \omega M I_2 \tag{44}$$

The primary current and reactance can be found by writing down a similar equation for the secondary circuit and solving. This is not necessary, however, as the solution already obtained for direct coupling applies to this case also. Consider L_1 to be made up of two inductances in series, M and L_a, the latter being given by $L_a = L_1 - M$. Similarly consider L_2 to consist of two parts in series, M and $L_b = L_2 - M$. Then Fig 38 is replaced by Fig. 36, and equation (44) becomes

$$E = \left(\omega L_a + \omega M - \frac{1}{\omega C_1}\right) I_1 - \omega M I_2 \tag{45}$$

or,

$$E = \left(\omega L_a - \frac{1}{\omega C_1}\right) I_1 + \omega M (I_1 - I_2)$$

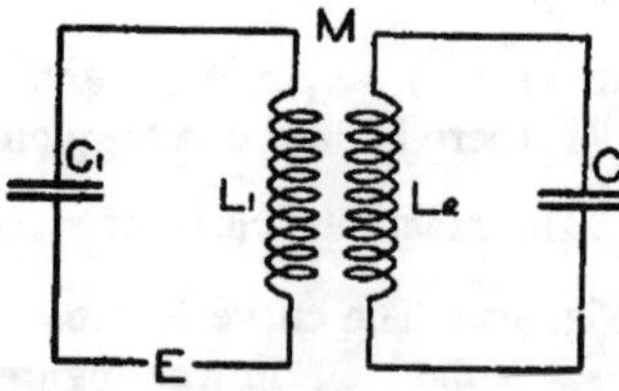

Fig. 38.—*Circuits involving inductive coupling*

This is the same as equation (32) for direct-coupled circuits. The two cases are, therefore, equivalent.

Equivalent Direct Coupling.—Thus, an inductively coupled system may be considered to be replaced by the direct-coupled system of Fig. 36, in which

$$L_a = L_1 - M$$

$$L_b = L_2 - M$$

The reactance curves are the same, and the frequencies of maximum current, given by ω' and ω'', (35) and (36), are the same. Equations (34) are more convenient in the following form:

$$k = \frac{M}{\sqrt{L_1 L_2}}, \quad \omega_1 = \frac{1}{\sqrt{L_1 C_1}}, \quad \omega_2 = \frac{1}{\sqrt{L_2 C_2}}$$

The example given in Fig. 37 was actually for direct-coupled circuits, but corresponds also to a case of inductively coupled circuits in which

$$L_1 = 297 \times 10^{-6} \text{ henry,}$$

$$L_2 = 450 \times 10^{-6} \text{ henry,}$$

and C_1, C_2, and M are the same as before.

The special cases treated above, in which $\omega_1 = \omega_2$ may be considered as special cases of inductive coupling as well as of direct coupling, except that the last case, where $L_1 = L_2$, is of no particular interest when the coupling is inductive, because when M is varied L_a is not usually kept constant. With inductive coupling, M is usually varied by moving the coils with reference to one another, L_1 and L_2 remaining constant.

Example.—A test of this theory of inductively coupled circuits was made by a set of measurements upon two circuits arranged as in Fig. 38. The coupling was varied by changing the distance apart of the two coils L_1 and L_2. The effect of varying coupling is shown in Fig. 39. As the coils are brought closer together, increasing the coupling, the resonance points ω' and ω'' become more widely separated. The constants were as follows:

$C_1 = 0.000244$ microfarad,
$C_2 = 0.000098$ microfarad,
$L_1 = 103.5$ microhenries,
$L_2 = 246.9$ microhenries,
$M = 0.6, 2.0, 5.1, 25.0$ microhenries, successively.

The reactance curves were calculated from these data and the preceding formulas. The curves of current squared were plotted from observations of deflections of a galvanometer connected

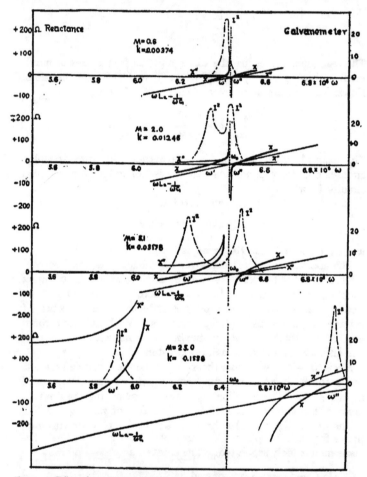

Fig. 39.—*Effect of varying the coupling upon reactance and resonance curves for inductively coupled circuits*

to a thermocouple loosely coupled to the secondary circuit, as a function of the frequency of the current which was induced in the primary by coupling loosely to a pliotron circuit. Each mutual inductance was measured by two measurements of the self-inductance of the two coils connected in series, the connections of one coil being reversed for the second measurement. While there are slight discrepancies in the agreement between the points of zero reactance and maximum current, due to slight changes of the inductances with frequency, the agreement is considered very good.

Effect of Coupling on Currents.—To calculate the current in the coupled circuits requires that account be taken of the resistances. A specially important case is that in which the primary and secondary circuits are both tuned, so as to be separately in resonance with the applied electromotive force,

$$\omega = \frac{1}{\sqrt{L_1 C_1}} = \frac{1}{\sqrt{L_2 C_2}} \tag{46}$$

Letting R_1 = resistance of primary circuit and R_2 = resistance of secondary, it may be shown [4] that

$$I_1 = \frac{E R_2}{R_1 R_2 + \omega^2 M^2}$$

$$I_2 = \frac{E \omega M}{R_1 R_2 + \omega^2 M^2}$$

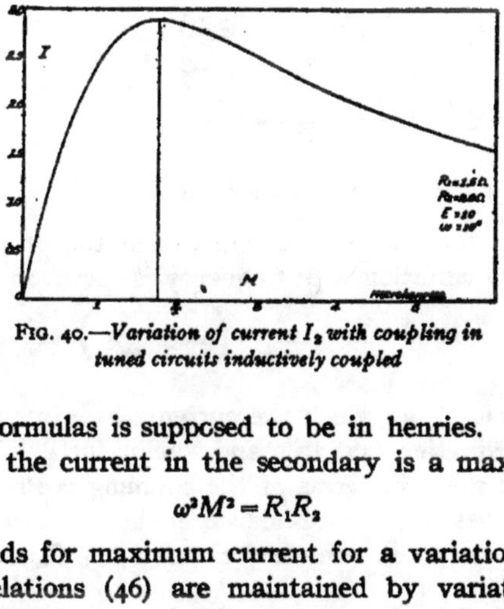

Fig. 40.—*Variation of current I_2 with coupling in tuned circuits inductively coupled*

M in these formulas is supposed to be in henries. For varying values of M the current in the secondary is a maximum when

$$\omega^2 M^2 = R_1 R_2$$

This also holds for maximum current for a variation of ω, provided the relations (46) are maintained by variations of the capacities. Other cases of this sort are solved in the reference cited below.[4]

[4] See reference Nos. 15 and 24, Appendix 2.

18. CAPACITIVE COUPLING

The phenomena in a pair of coupled circuits joined by capacitive coupling may be shown in a manner similar to the above discussion of direct coupling. Denoting by I_1 and I_2 the currents in the primary and secondary, respectively, the current in C_m is $I_1 - I_2$, and

$$E = \left(\omega L_1 - \frac{1}{\omega C_a}\right) I_1 + \left(\omega L_2 - \frac{1}{\omega C_b}\right) I_2$$

$$\left(\omega L_2 - \frac{1}{\omega C_b}\right) I_2 = -\frac{1}{\omega C_m}(I_1 - I_2) = -\frac{\dfrac{1}{\omega C_m}\left(\omega L_2 - \frac{1}{\omega C_b}\right)}{\omega L_2 - \frac{1}{\omega C_b} - \frac{1}{\omega C_m}} I_1$$

Therefore,

$$\frac{E}{I_1} = \omega L_1 - \frac{1}{\omega C_a} - \frac{\dfrac{1}{\omega C_m}\left(\omega L_2 - \frac{1}{\omega C_b}\right)}{\omega L_2 - \frac{1}{\omega C_b} - \frac{1}{\omega C_m}} \tag{47}$$

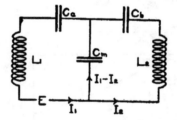

Fig. 41.—*Circuits involving capacitive coupling*

This is the expression for reactance to current in the primary circuit. A curve of its variation with frequency is shown in Fig. 42. The total reactance is X, the sum of $\left(\omega L_1 - \frac{1}{\omega C_a}\right)$ and X'', the last term in (47).

The values of ω' and ω'' at which the currents have maxima are readily found by equating (47) to 0 and solving for ω. It is convenient to express them in terms of the coupling coefficient and the respective values of ω for resonance in the primary circuit $L_1 C_a C_m$ alone and in the secondary circuit $L_2 C_b C_m$ alone. Using the relations,

$$k = \sqrt{\frac{C_a C_b}{(C_a + C_m)(C_b + C_m)}}, \quad \omega_1 = \sqrt{\frac{C_a + C_m}{L_1 C_a C_m}}, \quad \omega_2 = \sqrt{\frac{C_b + C_m}{L_2 C_b C_m}},$$

it turns out that the currents have maxima at the two frequencies given by

$$\omega' = \sqrt{\frac{\omega_1{}^2 + \omega_2{}^2 + \sqrt{(\omega_1{}^2 - \omega_2{}^2)^2 + 4\ k^2\omega_1{}^2\omega_2{}^2}}{2}} \qquad (48)$$

$$\omega'' = \sqrt{\frac{\omega_1{}^2 + \omega_2{}^2 - \sqrt{(\omega_1{}^2 - \omega_2{}^2)^2 + 4\ k^2\omega_1{}^2\omega_2{}^2}}{2}} \qquad (49)$$

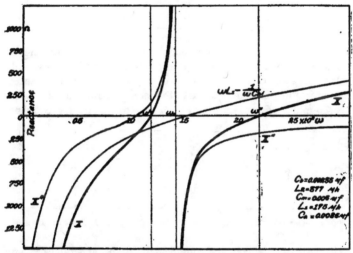

Fɪɢ. 42.—*Reactance diagram for capacitive coupling shown in Fig. 41*

Special Cases.—When $\omega_1 = \omega_2$, these expressions simplify to

$$\omega' = \omega_1\sqrt{1 + k}$$

$$\omega'' = \omega_1\sqrt{1 - k}$$

When the coupling is very loose, k approaches o, and

$$\omega' = \omega'' = \omega_1$$

The system responds simply to the frequency of resonance of either circuit by itself. When, on the other hand, C_m is small in comparison with C_a and C_b, the coupling is close and in the limit (48) and (49) reduce to

$$\omega' = \sqrt{2}\ \omega_1$$

$$\omega'' = 0$$

Practically this means that the system responds to only one frequency, given by $\omega' = \sqrt{\dfrac{2}{L_1 C_m}}$. It should be noted that ω' is greater than ω_1, while in the similar case of direct coupling ω' is less than ω_1.

When $\omega_1 = \omega_2$ and $L_1 = L_2$ (of course also $C_a = C_b$),

$$\omega' = \sqrt{\frac{2C_a + C_m}{L_1 C_a C_m}}$$

$$\omega'' = \sqrt{\frac{1}{L_1 C_a}}$$

As in the similar case of direct coupling, one of the frequencies is constant, not varying when C_a is kept constant and C_m is varied.

More General Cases.—The kind of capacitive coupling treated in the foregoing is a simple case of the more general type of capacitive coupling shown in Fig. 43. The expressions for coupling

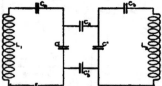
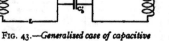
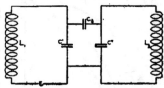

FIG. 43.—*Generalized case of capacitive coupling*

FIG. 44.—*Special type of case shown in Fig. 43*

coefficient, etc., which are complicated in the general case, are treated by E. Bellini (La Lumière Electrique, **32**, p. 241; 1916). Another simple case which has been found useful is that shown in Fig. 44. For this kind of capacitive coupling,

$$k = \frac{C_s}{\sqrt{(C' + C_s)(C'' + C_s)}}$$

Here again, in the special case of $\omega_1 = \omega_2$ and $C' = C''$, one of the frequencies is constant, not varying when C' is kept constant and C_s is varied.

For further information on coupled circuits, calculation of the currents, transformation ratios, etc., the reader is referred to Fleming's The Principles of Electric Wave Telegraphy and Telephony, Chapter III.

19. CAPACITY OF INDUCTANCE COILS

The small capacities between the turns of a coil are of such importance in radio design and measurements that a coil can seldom be regarded as a pure inductance. The effect of this distributed capacity is ordinarily negligible at low frequencies, but it modifies greatly the behavior of a coil at radio frequencies. For most purposes a coil can be considered as an inductance with a small capacity in parallel as shown in Fig. 45. This fictitious equiva-

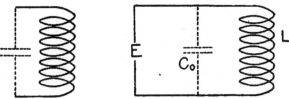

FIG. 45.—*Circuit which is equivalent to a*   FIG. 46.—*Coil having capacity, with emf in*
coil having distributed capacity   *series; a case of parallel resonance*

lent capacity is called the capacity of the coil. Investigations have shown that in ordinary coils its magnitude does not vary with frequency. Thus a coil may in itself constitute a complete oscillating circuit even when the ends of the coil are open.

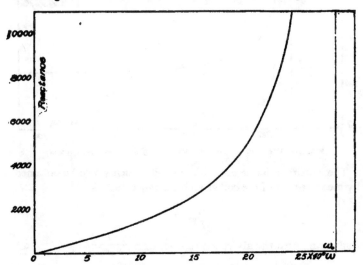

FIG. 47.—*Reactance diagram for coil having capacity with emf in series*

Emf in Series with the Coil.—If such a coil is placed in a circuit with an electromotive force in series, the case is one of parallel resonance. The reactance curve will be as shown in Fig. 47, which has the same shape as the left branch of the resultant in

35601°—18——5

Fig. 28. The right branch is of no interest and is not shown here, because for higher frequencies than ω_0 (at which the reactance becomes infinite) the coil no longer functions as an inductance. If the resistance is negligible, the current due to the electromotive force E is

$$I = \left(\frac{1}{\omega L} - \omega C_0\right) E$$

$$\therefore \frac{E}{I} = \frac{1}{\dfrac{1}{\omega L} - \omega C_0} = \frac{\omega L}{1 - \omega^2 C_0 L}$$

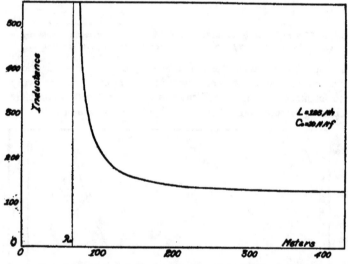

FIG. 48.—*Variation of apparent inductance of a coil with wave length*

The apparent inductance of the coil, which would be obtained by measurement of the coil as an inductance, is L_a in

$$\frac{E}{I} = \omega L_a$$

Comparing with the above expression, $L_a = \dfrac{L}{1 + \omega^2 C_0 L}$ (50)

When $\omega^2 C_0 L$ is small compared with 1, this becomes

$$L_a = L(1 + \omega^2 C_0 L),$$ (51)

at frequencies remote from ω_0, and for C_0 in farads and L in henries.

It is usually convenient to calculate the apparent inductance in terms of wave length. (See sec. 78.) Equation (51) becomes

$$L_a = L\left(1 + 3.553\,\frac{C_0 L}{\lambda^2}\right),$$

where C_0 is in micromicrofarads, L in microhenries, and λ in meters. This holds except for wave lengths near that corresponding to ω_0, in which case the more accurate expression applies:

$$L_a = \frac{L}{1 - 3.553\,\dfrac{C_0 L}{\lambda^2}} \tag{52}$$

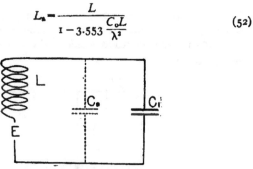

Fig. 49—*Coil having capacity with emf generated in the coil*

As λ_0, the wave length corresponding to ω_0, is approached, L_a becomes very large. λ_0 is the wave length at which the inductance and capacity of the coil would be in resonance, the coil itself constituting a complete oscillating circuit. This is a good example of parallel resonance. The wave length λ_0 is called the fundamental wave length of the coil, similar to the fundamental wave length of an antenna.

Emf Induced in Coil.—If a condenser is connected across the ends of a coil and the coil is loosely coupled to a source so that an electromotive force is induced in the coil, the total capacity in the circuit will be the sum of the condenser capacity and the coil capacity. This is shown in Fig. 49, C_0 and C_1 being in parallel, and the induced emf being indicated by the electromotive force E. If now the inductance be calculated for any wave length from the capacity of the condenser C, which causes resonance at that wave length, taking no account of the coil capacity C_0, the apparent inductance L_a so obtained will be greater than the pure inductance L. This is readily seen from

$$\omega^2 = \frac{1}{L_a C_1}$$

$$\omega^2 = \frac{1}{L(C_0 + C_1)}$$

$$\frac{L_a}{L} = \frac{C_0 + C_1}{C_1}$$

$$L_a = L\left(1 + \frac{C_0}{C_1}\right) \tag{53}$$

Thus the apparent inductance becomes greater the smaller the capacity C_1 connected to the coil to produce resonance—i. e., the smaller the wave length at which the measurement is made—just as in Fig. 48 above.

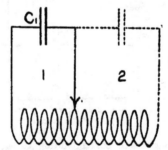

FIG. 50.—*Effect of distributed capacity in the unused turns of a coil*

Formula (53) is identically equivalent to formula (50) above. Thus the apparent inductance of a coil varies with the frequency in the same manner whether the electromotive force is applied in series with the coil or by induction in the coil itself. The pure induction L and the capacity C_0, to which the coil is equivalent, may be determined by either of the two methods given on page 136 below. The simple theory applies when the resistance is negligible, a condition which is ordinarily met at radio frequencies.

Effects of Dead Ends.—The capacities of coils frequently give rise to peculiar and undesirable effects in radio circuits. Among these are the effects caused by the capacities of those parts of a coil which are not connected in the circuit. For example, in many radio sets spiral coils with many turns are used in which the inductance may be varied by attaching a clip to any turn, thus utilizing more or fewer turns. The turns which are supposedly "dead" may actually produce considerable effect upon the circuit both in respect to energy loss and the frequency of resonance. As in the diagram (Fig. 50), suppose that a few turns of the coil are connected across a condenser C_1 (circuit 1) and high frequency oscillations are set up in this circuit. The numerous overhanging or unused turns are in the magnetic field of the used turns and are closed by their capacity (indicated by dotted lines). Hence this sec-

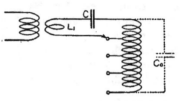

FIG. 51.—*Wavemeter circuit having coil with taps*

ond circuit is coupled closely to the first and if the resonant frequency of this circuit is near that of the first, considerable current will flow in it, strongly affecting the apparent resistance and resonant frequency of the first circuit. Indeed, the circuit 1 may, in consequence, respond to two frequencies, as in previously considered cases of coupled circuits. The case is the same as that of Fig. 31, and the reactance and current curves of the system would be of the same form as in Fig. 32. In this case of many overhanging turns compared to the number in use, it is generally advisable to short-circuit the overhanging turns, for then the impedance of the second circuit becomes so high that little current will flow. When only a few turns are overhanging it is best to leave them open, for the overhanging turns will have a greater impedance when open than when short-circuited.

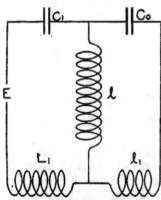

FIG. 52.—*The case of coupled circuits to which Fig. 51 is equivalent*

Another common case in which dead ends cause troublesome effects is shown in Fig. 51, a common form of circuit used in receiving apparatus and in some forms of wavemeters. L' is a coupling coil. The main coil L, of large inductance and considerable distributed capacity, is divided into sections, one or more of which may be connected in circuit to allow adjustment for various ranges of wave lengths. The fictitious condenser C_0 in parallel with this coil represents the effective capacity of the coil. For short wave lengths only part of the coil is in circuit, the unused sections being, however, inductively related to the part in circuit. It will be seen that this arrangement is really a case of two direct-coupled circuits, which should respond to two distinct frequencies or wave lengths. The circuit may be diagrammatically represented by the circuit in Fig. 52, which is equivalent to Fig. 31 above. This circuit is resonant to two frequencies, as shown by the resultant reactance curve of Fig. 32. As the setting

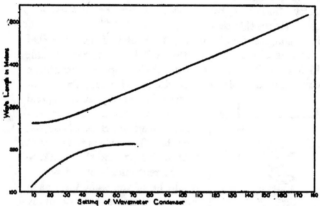

Fig. 53.—*Calibration curve of a commercial wave meter with discontinuity caused by distributed capacity*

of the condenser C_1 is varied, the frequencies to which the circuit responds are changed. For settings in the neighborhood of the natural frequency of the circuit $ll'C_0$ strong currents are obtained at both the resonant frequencies. For frequencies considerably more remote, the current of only one frequency is appreciable.

This behavior is sometimes experimentally found in wave meters. As the setting of the condenser is varied, the frequencies or wave lengths to which the current responds vary, and in the neighborhood of a certain wave length there are two wave lengths at which resonance occurs for every condenser setting. This is shown in Fig. 53, which is an experimentally obtained calibration curve of a commercial wave meter.

20. THE SIMPLE ANTENNA

Distributed Capacity and Inductance.—The current flowing into a condenser is given by $I = E\omega C$, and the voltage across an inductance is given by $E = I\omega L$. Thus the current into a condenser (voltage constant) and the voltage across an inductance (current constant) increase as the frequency increases. Both of these facts tend to make the small capacities between different portions of a circuit more important, the higher the frequency. At low frequencies in general the current at different points in a circuit is the same, and displacement currents are present only where relatively large condensers have been intentionally inserted in the circuit. The inductance and capacity are definitely localized or "lumped." At very high frequencies, however, or when the dimensions of the circuit are comparable to the wave length, the

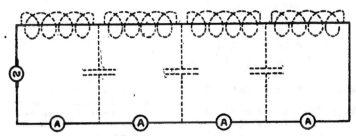

Fig. 54.—*Circuit representing distributed capacity and inductance*

capacities between different parts of the circuit become important and the current may vary appreciably from point to point in the circuit. Some of the current leaks away from or onto the conductor through the capacity to other parts of the circuit, the current through inductances in different parts of the same circuit will be different, and hence their inductive effect will be different. In such a case the equivalent capacity and inductance of the circuit will depend upon the frequency and the separate condensers and inductances must be considered with regard to their position in the circuit—i. e., one has to deal with "distributed" inductance and capacity. Fig. 54 represents a circuit or line of two long parallel wires supplied with current from a generator and closed at the far end. The inductance of the wires and the capacity between them are represented by condensers and inductances drawn in dotted lines. A number of ammeters are, for convenience, supposed to be inserted at points in the circuit. At a low frequency very little current will flow into the condensers and all of the ammeters will read the same. If the frequency is increased, more current will flow through the condensers and the ammeter readings will decrease successively from the generator to the far end. As a result of the changed distribution of current in the line, the equivalent inductance, capacity, and resistance of the line will vary with the frequency.

Simple Antenna.—The simplest form of antenna is a single vertical wire, the lower end of which is connected to ground. This forms an oscillatory circuit, the inductance is due to the wire and the capacity is that between the wire and the ground. Thus Fig. 55 shows diagrammatically the capacity and inductance and the flow of current at an instant of time. Some of the current from the wire is continually flowing off by the capacity paths to ground,

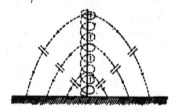

Fig. 55.—*Distributed capacity and inductance in a vertical wire, a simple form of antenna*

Fig. 56.—*Distribution of current and voltage in the simple antenna when oscillating at its fundamental frequency*

so that the maximum current is flowing at the base of the antenna, while at the extreme top there is no current flowing in the wire. The amplitude of the voltage alternations is zero at the ground and is a maximum at the top. The distribution of current and voltage in approximately sinusoidal and is shown in Fig. 56. This represents the fundamental oscillation of a simple antenna. The length of the wire is equal to the distance from node to loop or is one-fourth of the wave length. It is possible for the simple antenna to oscillate with other distributions of current and voltage, in which, however, the top must always be a node for current and the bottom a node for voltage. Thus in Fig. 57 is shown the next possible oscillation. Here the length of the wire is three-fourths of a wave length. Hence the wave length is one-third of the fundamental or the frequency is three times as great. Other possible oscillations have frequencies of five, seven, nine, etc., times the fundamental.

Antenna with Large End Capacity.—Suppose that a number of long horizontal wires are attached to the top of the vertical wire of the simple antenna, thus forming an inverted "L" antenna as in Fig. 58. In this case only a small proportion of the current in the vertical portion flows off to ground through capacity paths, the main capacity flow taking place from the horizontal portion. Thus the current throughout the vertical portion will be very nearly constant. The total capacity will be much larger than that of the simple antenna and the inductance likewise somewhat larger, hence the wave length will be considerably increased.

There are a number of other forms of antennas which also have large capacity areas at the top of the vertical lead, such as the "T," "umbrella," etc.

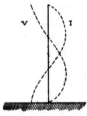

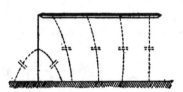

FIG. 57.—*The distribution of current and voltage in the simple antenna when oscillating in the first harmonic*

FIG. 58.—*Form of antenna of large capacity*

21. ANTENNA WITH UNIFORMLY DISTRIBUTED CAPACITY AND INDUCTANCE

The mathematical treatment of currents in circuits having distributed capacity and inductance are generally concerned with the case where these quantities are uniformly distributed. Because of end effects, this condition can not be strictly realized except with circuits of infinite length, such as two parallel wires, a single wire with a concentric cylindrical return, or a single wire (or number of parallel wires), and ground return. Such theory applies approximately, however, to the simple vertical-wire antenna or to the horizontal portion of an inverted "L" antenna. The following notation is used in this discussion:

l=length of antenna, CD. (Fig. 59),
L_1=inductance per unit length.
C_1=capacity per unit length.
$L_0=lL_1$=inductance for uniform current.
$C_0=lC_1$=capacity for uniform voltage.
L_a=low-frequency inductance of antenna=$1/3\ L_0$.
C_a=low-frequency capacity of antenna=C_0.
X=reactance of antenna.
X_L=reactance of loading coil.
X_C=reactance of series condenser.
L=loading coil.
C=series condenser.

In Fig. 59 an inverted "L" antenna is drawn to represent a circuit with uniformly distributed capacity and inductance, the distributed quantities being represented by dotted lines. The resistance is assumed to be negligible. CD is the horizontal portion, BE the ground, and BC the lead-in which is supposed to be free from inductance or capacity, excepting when a coil or condenser or both are inserted at A.

Low-Frequency Capacity and Inductance of Antenna.—It is desirable to explain the significance of the quantities L_1 and C_1, and the quantities $C_0 = lC_1$ and $L_0 = lL_1$. If the portion CD were uniformly charged to unit positive potential, then the charge on each unit of length of CD would be numerically equal to C_1 and the total charge on CD would be $C_0 = lC_1$. The antenna would be charged in this way if a constant or slowly alternating emf were introduced at A, and hence the quantity $C_0 = lC_1$ is sometimes

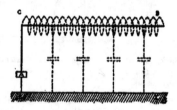

FIG. 59.—*Antenna with uniformly distributed capacity and inductance*

called the "static" capacity of the antenna. It is also called the low-frequency capacity. Formulas for the calculation of the static capacity are given on pages 237 to 242. At high frequencies the potential at all points on CD is, however, not the same at a given instant, and hence, the capacity (called the high-frequency or dynamic capacity) is different. In the case of the inductance, suppose that a conductor of negligible resistance and inductance is connected from D to E and that an emf which is constant or slowly alternating is introduced at A. The current flow at all portions of the circuit $CDEB$ at a given instant would then be the same and the total inductance L_0 of the circuit would be numerically equal to the total magnetic flux of lines linked with the circuit when the current is unity. This is the value of inductance which would be calculated from the formulas on pages 247 to 250 The inductance per unit length L_1 is L_0 divided by the length. In the actual antenna, however, the current flow can not be the same at all portions of the circuit, for the current at the open end of the aerial must always be zero, and hence $L_0 = lL_1$ does not represent the inductance of the aerial at any frequency. It will now be shown that the low-frequency inductance of the aerial is one-third L_0.

If an alternating voltage of frequency f (or $\omega = 2\pi f$) is introduced at A in the actual antenna, it can be shown[5] that the aerial or horizontal portion behaves as a reactance,

$$X = -\sqrt{\frac{L_1}{C_1}} \cot \omega l \sqrt{C_1 L_1},$$

or,

$$X = -\sqrt{\frac{L_0}{C_0}} \cot \omega \sqrt{C_0 L_0}$$

The expansion of the cotangent of an angle into a series is

$$\cot x = \frac{1}{x} - \frac{x}{3} \cdots \cdots$$

the remaining terms being negligible if x is small. Hence, if ω is small—i. e., for low frequencies—substituting $\omega \sqrt{C_0 L_0}$ for x in the series, we have

$$X = -\sqrt{\frac{L_0}{C_0}} \left(\frac{1}{\omega \sqrt{C_0 L_0}} - \frac{\omega \sqrt{C_0 L_0}}{3} \cdots \cdots \right) = -\frac{1}{\omega C_0} + \frac{\omega L_0}{3}$$

This is the reactance of a capacity C_0 in series with an inductance $\dfrac{L_0}{3}$, as in Fig. 60; hence we see that the static capacity is $C_0 = lC_1$ and the low-frequency inductance is $\dfrac{L_0}{3} = \dfrac{lL_1}{3}$.

FIG. 60.—*Simple circuit which is equivalent to the antenna shown in Fig. 59*

Fundamental Wave Length.—The expression for the reactance of the aerial

$$Z_1 = -\sqrt{\frac{L_0}{C_0}} \cot \omega \sqrt{C_0 L_0}$$ shows that at low frequencies when the cotangent is positive, the reactance is negative, and hence the capacity reactance overbalances the inductive. As the frequency increases the cotangent becomes zero, which first occurs when $\omega \sqrt{C_0 L_0} = \dfrac{\pi}{2}$ and the reactance is zero. The frequency at which this occurs is the fundamental frequency of natural oscillation of

[5] See reference No. 31, Appendix 2.

the antenna when the vertical lead-in is free from inductance and capacity.   This frequency is

$$f = \frac{\omega}{2\pi} = \frac{1}{4\sqrt{C_o L_o}}.$$

The fundamental wave length is given by

$$\lambda = \frac{c}{f} = 4c\sqrt{C_o L_o}$$

$$\lambda_m = 1199.\sqrt{C_o L_o} \qquad (54)$$

where c = velocity of propagation of electromagnetic waves, and in the last equation λ_m stands for wave length in meters, C_o is in microfarads, and L_o in microhenries.

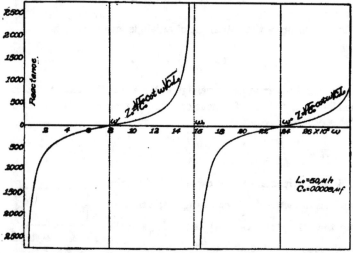

FIG. 61.—*Reactance curve for antenna with uniformly distributed capacity and inductance* (Z on curves corresponds to X in text)

Harmonics.—For frequencies above the fundamental the reactance becomes positive and the inductive reactance preponderates up to the frequency $f = \frac{1}{2\sqrt{C_o L_o}}$ at which the reactance becomes infinite.   Beyond this frequency the reactance is again negative, but decreases numerically with increasing frequency until it again becomes zero and there is a harmonic natural oscillation at a frequency

$$f = \frac{3}{4\sqrt{C_o L_o}}$$

This variation of the aerial reactance with the frequency is shown by the cotangent curves in Fig. 61. Those frequencies for which the reactance of the aerial becomes zero and which are the natural frequencies of oscillation of the antenna when the lead-in is of zero reactance are given in the figure by the points of intersection of the cotangent curves with the axis of ordinates. For these points

$$f = \frac{m}{4\sqrt{C_o L_o}}; \quad m = 1, 3, 5, \text{ etc.}$$

$$\lambda = \frac{4c}{m} \sqrt{C_o L_o}.$$

Now it is shown in the theory of long circuits with uniformly distributed quantities, such as those considered here, that at high frequencies $\sqrt{C_1 L_1} = \frac{1}{c}$, approximately, and since $\sqrt{C_o L_o} = l\sqrt{C_1 L_1}$, it follows that

$$f = \frac{mc}{4l}$$

$$\lambda = \frac{4l}{m}$$

Hence the wave lengths of the fundamental and harmonic oscillations of the antenna are approximately $4l$, $4/3l$, $4/5l$, etc., or, as stated above in the description of the oscillations of a simple vertical antenna, the length of the wire is approximately $1/4$, $3/4$, $5/4$, etc., times the wave length. The distributions of current and voltage corresponding to these different possible modes of oscillation have also been shown. These relations between antenna length and wave length are only rough approximations, because of the finite length of the antenna and because the vertical portion of the antenna has been neglected.

22. LOADED ANTENNA

In general, when an antenna is used for transmission or reception, coils or condensers or both are inserted in the lead-in to modify the natural frequency of oscillation of the system; i. e., to tune it to a given frequency or wave length. When a reactance X_x is present in the lead-in, the natural frequency of oscillation is then determined by the condition that the total reactance of lead-in plus aerial shall be zero—that is, $X_x + X = 0$. In the succeeding sections the cases will be considered where an inductance or a condenser is inserted in the lead-in.

Antenna with Series Inductance.—If an inductance L_0 is inserted in the lead-in, its reactance X_L is equal to ωL. This is a positive reactance which increases linearly with the frequency and is represented in Fig. 62 by a straight line. The reactance of the aerial X is shown by the cotangent curves. The sum $X_L + X$ is drawn in heavy solid lines. Those frequencies at which these latter curves cut the axis of abscissas are the frequencies for which

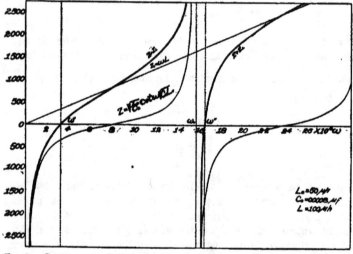

Fig. 62.—*Reactance curve for antenna with series inductance.*   (*Z on curves corresponds to X in text*)

$X_L + X = 0$ and are the natural frequencies of oscillation of the system. It will be noted that the insertion of the inductance coil has decreased the natural frequencies of oscillation—i. e., increased the wave length. Also the harmonic frequencies are no longer integral multiples of the fundamental as in the case of the simple antenna.

The condition $X_L + X = 0$, which determines the natural frequencies of oscillation, leads to the equation

$$\omega L - \sqrt{\frac{L_0}{C_0}}\,\cot \omega\sqrt{C_0 L_0} = 0$$

or

$$\frac{\cot \omega\sqrt{C_0 L_0}}{\omega\sqrt{C_0 L_0}} = \frac{L}{L_0}.$$

Since $\sqrt{C_0 L_0} = \frac{l}{c}$; $\omega = 2\pi f$ and $\lambda = \frac{c}{f}$ this equation may also be written

$$\frac{\cot \frac{2\pi l}{\lambda}}{\frac{2\pi l}{\lambda}} = \frac{L}{L_0}.$$

These transcendental equations which determine ω and λ can not be solved directly; it is, however, possible to solve them graphically as shown in Fig. 62 or to determine indirectly a table (second column of Table 1) which will give the values of $\omega\sqrt{C_0 L_0}$ or $\frac{2\pi l}{\lambda}$ for different values of $\frac{L}{L_0}$, from which then ω, f, or λ may be determined.

Table 1 gives only the lowest value of $\omega\sqrt{C_0 L_0}$ or $\frac{2\pi l}{\lambda}$, corresponding to the fundamental oscillation of the loaded antenna. In any actual antenna the wave length would be greater than that given by this calculation because of the inductance and capacity of the vertical portion; this discussion deals only with the horizontal portion.

As an example of the method let us assume the quantities used in Figs. 61 and 62. Let the length of the antenna be $l = 60$ meters and the static capacity $C_0 = lC_1 = 0.0008$ microfarad. Then since

$$\sqrt{L_0 C_0} = \frac{l}{c} = \frac{60}{3 \times 10^8}, \quad L_0 = 50 \text{ microhenries}.$$

In the case of the unloaded antenna, the natural wave lengths would be $4l$, $4/3l$, etc.—that is, 240, 80, etc., meters; the natural frequencies $\left(f = \frac{c = 3 \times 10^8}{\lambda \text{ meters}} \right)$ would be 1.25×10^6, 3.75×10^6, etc., cycles per second; the periodicities $(\omega = 2\pi f)$ 7.85×10^6, 23.6×10^6, etc., radians per second agreeing with the values in Fig. 62.

If now an inductance $L = 100$ microhenries is introduced in the lead-in, we have $\frac{L}{L_0} = 2$. From Table 1 we find that

$$\omega\sqrt{C_0 L_0} = \frac{2\pi l}{\lambda} = 0.653,$$

hence

$$\omega = \frac{0.653}{\sqrt{0.0008 \times 10^{-6} \times 50 \times 10^{-6}}} = 3.26 \times 10^6$$

and

$$\lambda = \frac{2\pi l}{0.653} = 577 \text{ meters}.$$

This corresponds to the lowest frequency of oscillation as shown in Fig. 62. Introducing the inductance has increased the wave length of this oscillation from 240 to 577 meters.

The harmonic oscillations are of importance in some cases. If the emf which is applied to the antenna has the fundamental frequency of the loaded antenna, the oscillations will be of that frequency alone. If, however, the antenna is first charged and then set into oscillation by the breaking down of a spark gap in the antenna as in the original Marconi antenna, frequencies corresponding to all of the possible modes of oscillation will be emitted

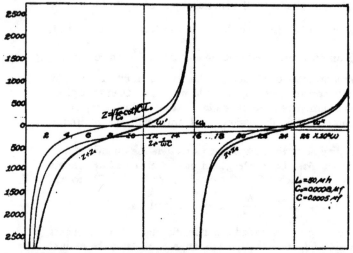

Fig. 63.—*Reactance curve for antenna with series condenser* (Z *on curves corresponds to* X *in text*)

by the antenna. Or in the case of the arc, which in itself generates fundamental and harmonic frequencies, if a harmonic of the arc coincides somewhat closely with one of the harmonic modes of vibration of the antenna, this oscillation will be strongly reinforced, a large amount of energy will be wasted, and interference will be caused.

Antenna with Series Condenser.—Though not as important practically as the case just considered, there are occasions when a condenser is inserted in the lead-in of an antenna to shorten the wave length. If its capacity is C, its reactance will be $X_C = -\dfrac{1}{\omega C}$. In Fig. 63 the cotangent curves representing the aerial reactance X are again shown as is also the parabola representing X_C. The sum $X_0 + X$ is drawn in heavy solid lines and crosses the axis at points corresponding to the natural frequencies of oscillation of the loaded antenna. It will be noted that the frequencies of oscillation are increased—i. e., the wave length shortened—by the insertion of the condenser and that the harmonic oscillations are not integral multiples of the fundamental. Fig. 63 shows that ω is increased from 7.85×10^6 to 10.96×10^6 or the wave length decreased from 240 to 172 meters by the insertion of a 0.0005 microfarad condenser.

Simple Calculation of the Wave Length of a Loaded Antenna.— The ordinary formula for the frequency of oscillation of circuits with lumped inductance and capacity may be applied to the antenna with distributed constants in the case of an inductance coil in the lead-in, and the error in computing the frequency or wave length will be small. The inductance and capacity of the aerial at all frequencies is supposed to be the same as the low-frequency values; i. e., $\dfrac{L_0}{3}$ and C_0. If the loading coil has an inductance L, the total inductance will be $L + \dfrac{L_0}{3}$. Hence

$$\omega = \frac{1}{\sqrt{\left(L + \dfrac{L_0}{3}\right) C_0}}$$

the frequency is

$$f = \frac{1}{2\pi \sqrt{\left(L + \dfrac{L_0}{3}\right) C_0}}$$

and the wave length

$$\lambda_m = 1884. \sqrt{\left(L + \dfrac{L_0}{3}\right) C_0} \tag{55}$$

where in the last equation λ_m means wave length in meters, inductance is expressed in microhenries, and the capacity in microfarads.

Applying this to the numerical example worked out above by the exact theory, in which

$$\left(L + \dfrac{L_0}{3}\right) = 116.67 \text{ microhenries}$$

$$C_0 = 0.0008 \text{ microfarads}$$

we have $\lambda = 575$ meters, which differs only one-third of 1 per cent from the value $\lambda = 577$ obtained before.

The magnitude of the errors in using the simple formula

$$\omega = \frac{1}{\sqrt{\left(L + \dfrac{L_o}{3}\right)C_o}}$$

is also shown in Table 1. In the second column, as pointed out before, are the values of $\omega\sqrt{C_oL_o}$ or $\dfrac{2\pi l}{\lambda}$ for different values of $\dfrac{L}{L_o}$ as computed from the exact cotangent formula. The simple formula gives

$$\omega\sqrt{C_oL_o} = \frac{1}{\sqrt{\dfrac{L}{L_o} + \dfrac{1}{3}}}$$

and in the third column are given for comparison the values of $\omega\sqrt{C_oL_o}$ computed on this basis. These values are too high; i. e., result in too high a value for the frequency or too low a value for the wave length. The per cent error is given in the last column. The maximum error is 10 per cent for $L = 0$—i. e., at the fundamental of the antenna—but the error rapidly decreases as L increases and is less than 1 per cent for L equal to or greater than L_o.

It has been stated in several publications that very large errors would result from applying the ordinary theory of circuits with lumped constants to the case of an antenna which has distributed quantities. This misconception has arisen because the quantity L_o which occurs in the formula for the distributed case was used for the inductance of the aerial in applying the formula for the case of lumped constants. We have pointed out that L_o could not be the inductance of the aerial at any frequency. When $\dfrac{L_o}{3}$, instead of L_o, is used the agreement is very close. In fact, since this error is usually less than 1 per cent, it is practically never worth while to use formulas based on the precise theory to calculate wave length, because of the uncertainty introduced by the vertical portion of the antenna. Equation (55) is therefore sufficient for all ordinary calculations.

TABLE 1.—Data for Loaded Antenna Calculations

$\frac{L}{L_0}$	$\omega\sqrt{C_0L_0}$	$\dfrac{1}{\sqrt{\frac{L}{L_0}+\frac{1}{3}}}$	Difference, per cent	$\frac{L}{L_0}$	$\omega\sqrt{C_0L_0}$	$\dfrac{1}{\sqrt{\frac{L}{L_0}+\frac{1}{3}}}$	Difference, per cent
0.0	1.571	1.732	10.3	3.1	0.539	0.540	0.1
.1	1.429	1.519	6.3	3.2	.532	.532	.1
.2	1.314	1.369	4.2	3.3	.524	.525	.1
.3	1.220	1.257	3.0	3.4	.517	.518	.1
.4	1.142	1.168	2.3	3.5	.510	.511	.1
.5	1.077	1.095	1.7	3.6	.504	.504	.0
.6	1.021	1.035	1.4	3.7	.4977	.4979	.0
.7	0.973	0.984	1.1	3.8	.4916	.4919	.0
.8	.931	.939	0.9	3.9	.4859	.4860	.0
.9	.894	.900	.7	4.0	.4801	.4804	.0
1.0	.860	.866	.7	4.5	.4548	.4549	.0
1.1	.831	.835	.5	5.0	.4330	.4330	.0
1.2	.804	.808	.5	5.5	.4141	.4141	....
1.3	.779	.782	.4	6.0	.3974	.3974	....
1.4	.757	.760	.4	6.5	.3826	.3826	....
1.5	.736	.739	.4	7.0	.3693	.3693	....
1.6	.717	.719	.3	7.5	.3574	.3574	....
1.7	.699	.701	.3	8.0	.3465	.3465	....
1.8	.683	.685	.3	8.5	.3366	.3366	....
1.9	.668	.670	.3	9.0	.3275	.3275	....
2.0	.653	.655	.3	9.5	.3189	.3189	....
2.1	.640	.641	.2	10.0	.3111	.3111	....
2.2	.627	.628	.2	11.0	.2972	.2972	....
2.3	.615	.616	.2	12.0	.2850	.2850	....
2.4	.604	.605	.2	13.0	.2741	.2741	....
2.5	.593	.594	.2	14.0	.2644	.2644	....
2.6	.583	.584	.2	15.0	.2556	.2556	....
2.7	.574	.574	.2	16.0	.2476	.2476	....
2.8	.564	.565	.1	17.0	.2402	.2402	....
2.9	.556	.556	.1	18.0	.2338	.2338	....
3.0	.547	.548	.1	19.0	.2277	.2277	....
				20.0	.2219	.2219	....

23. ANTENNA CONSTANTS

Antenna Resistance.—The power supplied to maintain oscillations in an antenna is dissipated in three ways: (1) Radiation; (2) heat, due to conductor resistance; (3) heat, due to dielectric absorption. (At high voltages there is a further power loss due to brush discharge; this will not be considered in the following.) The first of these represents the only useful dissipation of power since it is the power which travels out from the antenna in the form of the electromagnetic waves which transmit the radio signals. The amount of power radiated depends upon the form of the antenna, is proportional to the square of the current flowing at the current antinode of the antenna, and inversely proportional to the square of the wave length of the oscillation. Since the dissipation of power is proportional to the square of the current, it may be considered to be caused by an equivalent or effective resistance, which is called the radiation resistance of an antenna. Thus the radiation resistance of an antenna is that resistance which, if inserted at the antinode of current in the antenna would dissipate the same power as that radiated by the antenna. The radiation resistance varies with the wave length in the same way as the radiated power; i. e., inversely as the square of the wave length. Curve 1 of Fig. 64 represents the variation of this component of the resistance of an antenna.

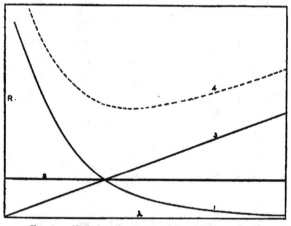

Fig. 64.—*Variation of antenna resistance with wave length*

The second source of dissipation of power, that due to ohmic resistance, includes the losses in the resistance of the wires, ground, etc., of the antenna. Due to eddy currents and skin effect in both the wires and ground, this resistance will vary somewhat with the wave length, being greater at shorter wave lengths. But in an actual antenna these changes are so small compared to other variations that we may regard this component of the total antenna resistance to be almost constant, as it is represented by the straight line 2 of Fig. 64. The third source of power dissipation—i. e., that due to dielectric absorption—is a result of the fact that the antenna capacity is an imperfect condenser. The magnitude of this power loss will depend upon the nature and position of imperfect dielectrics in the field of the antenna. Thus it has been found that a tree under an antenna may increase the resistance of the antenna enormously; buildings, wooden masts, and the antenna insulators also affect the absorption of the antenna capacity. It is pointed out in section 34 on con-

densers that the effective resistance of an absorbing condenser is proportional to the wave length. In the antenna, therefore, the loss of power due to dielectric absorption may be represented as taking place in a resistance which increases in proportion to the wave length. This component of the antenna resistance is represented in Fig. 64 by the straight line 3.

The curve of the total antenna resistance is obtained by combining these three resistance components as in curve 4 of the
same figure. This is the typical resistance curve of an antenna.
(See also Fig. 91, p. 126.) In the case of some antennas the resistance curve shows one or more humps at certain wave lengths.
This indicates the presence of circuits with natural periods of oscillation in the vicinity of the antenna, possibly the stays of the antenna, another antenna, or the metal structure of a building; and
the humps indicate that, at the particular wave length at which
they occur, these extraneous circuits are in tune with, and are absorbing power from, the antenna. The resistance curve of an antenna may be determined by several of the methods of resistance
measurement given in sections 47 to 50 below.

Measurement of Capacity and Inductance of an Antenna.—An
antenna is ordinarily used with a series loading coil. In the case
of uniform distribution of capacity and inductance, a formula and
table have been given (pp. 80 and 81) which permit the wave length
of resonance to be calculated for a given loading coil L when the
quantities C_0 and L_0 are known. C_0 is the low-frequency capacity
and $\frac{L_0}{3}$ the low-frequency inductance of the antenna. It has
furthermore been shown that the resonance wave length can be
calculated with sufficient accuracy from the simple formula applicable to a circuit with lumped capacity and inductance. The
capacity in the equivalent circuit is taken to be the low-frequency
capacity C_0 of the antenna, and the inductance is the sum of the
inductance of the loading coil L and the low-frequency inductance of the antenna $\frac{L_0}{3}$. Thus the low-frequency values of antenna capacity and inductance are sufficient for wave-length
calculations by either formula. These low-frequency values will
be called simply the capacity C_a and inductance L_a of the antenna.
In terms of the previous notation

$$C_a = C_0 = lC_1$$

$$L_a = \frac{L_0}{3} = \frac{lL_1}{3}$$

and the simple formula for the wave length becomes

$$\lambda_m = 1884\sqrt{(L+L_a)\,C_a} \qquad (56)$$

where inductance is in microhenries and capacity in microfarads.

Measurement by the Use of Two Loading Coils.—In order to determine C_a and L_a experimentally, two loading coils of different and known values are successively inserted in the antenna and the wave lengths determined for which the antenna is in resonance. This may be done as in Fig. 65, which shows the inserted inductance, S a source of oscillations, and W a stand-

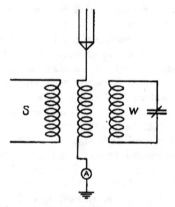

FIG. 65.—*Circuits for determining the capacity and inductance of an antenna*

ardized wave meter. The wave length of the source is varied until the antenna is in resonance as indicated by the ammeter or other indicating device on the antenna circuit. Then the antenna is detuned and the wave length of the source determined by the wave meter. Two coils L and L' are inserted and the corresponding wave lengths λ and λ' are determined. Using the simple equation (56) for lumped capacity and inductance

$$\left.\begin{array}{l} \lambda = 1884\sqrt{(L+L_a)\,C_a} \\ \lambda' = 1884\sqrt{(L'+L_a)\,C_a} \end{array}\right\} \qquad (57)$$

Eliminating C_a between these two equations and solving for L_a, we obtain

$$L_a = \frac{L'\lambda^2 - L\lambda'^2}{\lambda'^2 - \lambda^2} \qquad (58)$$

From the known values of L, L', λ, and λ' we obtain, therefore, the value of L_a and substituting this value in one of the original equations (preferably the one corresponding to the larger loading coil), we obtain the value of C_a.

Example.—An illustrative check upon this method will now be given. Let us suppose that the antenna has uniformly distributed capacity and inductance of certain values so that we can compute the wave lengths which would be observed by experiment when the loading coils L and L' are inserted. Then, from these wave lengths and the values of L and L' we will compute by the above formula (58) the values of L_a and C_a, and see how closely they agree with the original values $\dfrac{L_0}{3}$ and C_0. We will take the values $C_0 = 0.0008$ microfarad and $L_0 = 50$ microhenries used before, and a value for the first loading coil of $L = 100$ microhenries, for which we have previously found that the wave length would be 577 meters. In addition, we will take for the value of the second coil $L' = 400$ microhenries. From Table 1 we find for $\dfrac{L'}{L_0} = \dfrac{400}{50} = 8$, that $\omega \sqrt{C_0 L_0} = \dfrac{2\pi l}{\lambda'} = 0.346_5$ from which we get $\lambda' = \dfrac{2\pi l}{0.346_5} = 1088$ meters.

Substituting in formula (58) gives

$$L_a = \frac{400\,(577)^2 - 100\,(1088)^2}{(1088)^2 - (577)^2} = 17.3 \text{ microhenries.}$$

Using this value of L_a in equation (56), we have $1088 = 1884\sqrt{(400 + 17.3)C_a}$ from which

$$C_a = 0.000799.$$

But $\dfrac{L_0}{3} = 16.7$ microhenries, hence L_a differs from the correct value by 3.6 per cent. C_a differs from C_0 by only a little over a tenth of 1 per cent. These values are sufficiently accurate for most antenna measurements.

Corrected Values.—In measurements made with smaller values for one or both of the inserted coils, greater errors would arise, but these can be greatly reduced by a second approximation. Suppose that the inserted coils were of values $L = 50$ microhenries, $L' = 200$ microhenries. The values of λ and λ' would be found to be 438.4 and 785.2 meters. The value of L_a would come out 17.9 microhenries and C_a would equal 0.000797 microfarad, L_a differing from $\frac{L_o}{3}$ by 7.6 per cent. The errors are due to the fact that the simple formula does not hold exactly, the per cent error in L_a being magnified on account of the difference in the squares of the wave lengths in formula (58). However, this approximate value of L_a furnishes an approximate value of L_o and permits a close esti- mate of the error in the simple formula. These errors are given in Table 1 for different values of $\frac{L}{L_o}$. As shown there, the simple formula leads to low values for the wave length. If the observed wave lengths are corrected so as to correspond to the simple formula, accurate values of L_a and C_a will be obtained. Thus in the case of the 50-microhenry coil, since $L_o = 3L_a = 53.7$, approx- imately, the ratio $\frac{L}{L_o} = 0.9$ and from Table 1 the error is 0.7 per cent.

Reducing the observed wave length λ by this amount, we obtain $\lambda = 435.3$ meters. The correction in the case of the 200- microhenry coil is negligible, hence λ' is unchanged.

Recomputing, using $\lambda = 435.3$ and $\lambda' = 785.2$, we obtain

$$L_a = 16.6$$
$$\text{and } C_a = 0.000801$$

which are in satisfactory agreement with $\frac{L_o}{3}$ and C_o.

DAMPING

24. FREE OSCILLATIONS

Up to this point the high-frequency phenomena considered have been those produced when current of sine-wave form is used. Much of the theory remains substantially the same for the use of so-called damped waves. There are, however, certain important phenomena which are peculiar to damped waves and these will be treated in this section.

When a charged condenser is suddenly discharged in a circuit (Fig. 66) containing inductance and a moderate resistance in series with the capacity, an alternating current flows, but not of sine-wave form. If the resistance were zero, the current would have the sine-wave form and would continue to flow forever with undiminished amplitude. Every circuit contains resistance, how- ever, which means that the electric energy of the current is continually being converted into heat; and also in some circuits an appreciable amount of energy is radiated away from the circuit in electromagnetic waves. In consequence the amplitude of the alternating current in the circuit continually diminishes.

The behavior of such a circuit is similar to that of a compressed spring to which a weight is attached. When the spring is re-leased a vibration commences, which continues until the energy of the vibration has all been con-sumed by friction. If there is no external friction and the spring is very free from internal strains, etc., the vibration continues a very long time; but if the friction is great, the energy is all converted into heat in a short time. In the same way, the current in the circuit under consideration is reduced very rapidly if the resistance is large.

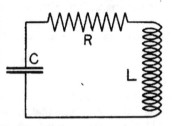

FIG. 66.—*Oscillatory circuit having resist-ance, inductance, and capacity in series*

A high-frequency current of continuously decreasing amplitude is called an "oscillatory" current, although the term "oscil-latory" is sometimes applied to any current of very high frequency. The decrease of amplitude is called "damping," and the current or wave is called a "damped" current or wave. (In contradistinction to a damped wave, a wave in which the amplitudes do not continuously decrease is called a persistent or sustained wave or oscillation.) The frequency of oscillation in a freely oscillating circuit depends only on the inductance and capac-ity of the circuit, if the resistance is not very large. The damping is determined by the resistance together with the inductance and capacity. The resistance is thus important in determining the character of the phenomena. Damped currents are in this respect distinct from those of the sine-wave form.

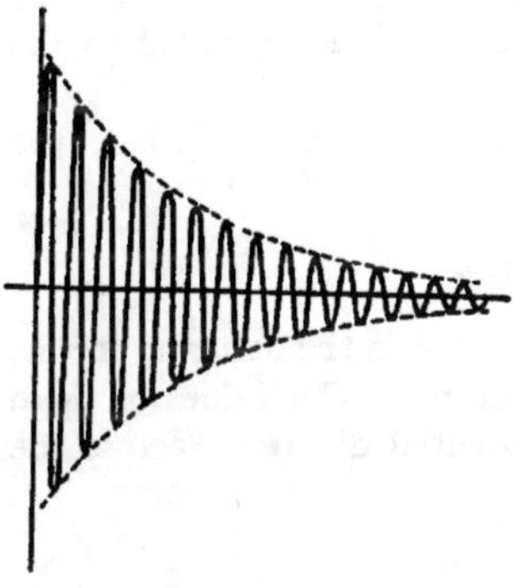

FIG. 67.—*Wave train with a logarithmic decrement of 0.2*

The oscillations which occur in a simple circuit upon which no external alternating emf is applied are called the "free" oscil-lations of the circuit. "Forced" oscillations, on the other hand, are those impressed on the circuit by an alternating emf from a source outside the circuit. When free oscillations are produced by the sudden discharge of a condenser, all of the energy which was stored in the condenser before discharge is lost from the circuit during the oscillations. The potential difference of the condenser, therefore, becomes lower and lower at every alternation of the current. Since there is no emf applied from outside the circuit, the potential differences of condenser, resistance, and inductance must balance, and their algebraic sum be zero.

$$\therefore L\frac{di}{dt} + Ri + \frac{\int i\,dt}{C} = 0 \tag{59}$$

This is the same as equation (18) given above in simple alternating-current theory, except that e, the applied emf, is here equal to 0. If the circuit contains a spark gap, the resistance R is not a constant, and the solution given immediately below does not apply.

Free oscillations may be produced by another method than the sudden discharge of the condenser in the circuit. If current is produced in the circuit by induction and the inducing emf is suddenly cut off, free oscillations are produced. The quenched gap is the means utilized in practice for suddenly cutting off the inducing emf, the gap being in a circuit closely coupled to the oscillating circuit.

The solution of equation (59) for any circuit in which the resistance is not extremely great, is

$$i = I_0 \epsilon^{-\alpha t} \sin \omega t$$

where I_0 is the initial current amplitude, α is the damping factor, and ω is 2π times frequency of oscillation. The values of these constants are, V_0 being the initial potential difference across the condenser,

$$I_0 = \omega C V_0$$

$$\alpha = \frac{R}{2L}$$

$$\omega = \frac{1}{\sqrt{LC}}$$

This last expression is an approximation for

$$\omega = \sqrt{\frac{1}{LC} - \frac{R^2}{4L^2}}.$$

The resistance of practically all high-frequency circuits is so small that the approximation is sufficiently accurate. Using these relations, it follows that

$$i = \sqrt{\frac{C}{L}}\, V_o \epsilon^{-\frac{R}{2L}t} \sin \frac{t}{\sqrt{LC}}.$$

The frequency of oscillation is called the "natural frequency" of the circuit. It should be noted that the natural frequency of a circuit in which the resistance is not very large is the same as the frequency of resonance to an electromotive force impressed upon the circuit. Thus it was shown in section 11 above, that when an alternating emf is impressed upon an inductance and capacity in series resonance occurs when $\omega L = \dfrac{1}{\omega C}$. This relation is the same as $\omega = \dfrac{1}{\sqrt{LC}}$, the expression above for the natural frequency.

Effective Value of Damped Current.—In the various spark types of radio transmitters the condenser is charged at regular intervals, each charge being followed by an oscillatory discharge. Fig. 68

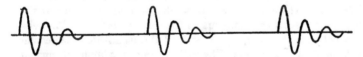

Fig. 68.—*Series of wave trains similar to that generated by a damped wave source*

roughly indicates the successive discharges; actual current curves are more complicated than this. The energy in each train of oscillations is practically all dissipated before the next train begins. Under these conditions the effective (root-mean-square) value of the current I in terms of the initial maximum current I_o is easily found, as follows (I=current which would be indicated by the steady deflection of an ammeter):

Let N = number of trains of oscillations per second.

W_1 = energy dissipated per train of waves.

W_2 = magnetic energy associated with current at beginning of each train.

Then
$$W_1 = \frac{RI^2}{N}$$

$$W_2 = \frac{1}{2}LI_o^2$$

These two energies must be equal, since it is assumed that the energy in each train of oscillations is dissipated before the next train begins.

$$\frac{RI^2}{N} = \tfrac{1}{2}LI_0^2$$

$$I^2 = \frac{NLI_0^2}{2R}$$

Since

$$\alpha = \frac{R}{2L},$$

$$I^2 = \frac{N}{4\alpha}I_0^2 \tag{60}$$

25. LOGARITHMIC DECREMENT

The rate of decrease of the current amplitudes during a train of waves is shown by the damping factor α defined above. When this is large, the curve ABC, drawn through the maxima of the oscillations in Fig. 67, approaches the axis quickly. The shape of the train of waves is such that the ratio of any maximum to the next one following is a constant. The rate of decrease of the amplitudes in a wave train is also indicated by the logarithmic decrement, which is defined to be the natural logarithm of the ratio of two successive maxima in the same direction.[6] The relation between the damping factor and the logarithmic decrement may be found as follows: The ratio of one maximum of current to the following is

$$\frac{I_0 \epsilon^{-\alpha t}}{I_0 \epsilon^{-\alpha(t+T)}}$$

where T is the period of oscillation and equals $\frac{2\pi}{\omega}$. This ratio $=$

$\epsilon^{\alpha T} = \epsilon^{\frac{2\pi\alpha}{\omega}}$. The napierian logarithm of the ratio is the logarithmic decrement δ which consequently equals $\frac{2\pi\alpha}{\omega}$. Since $\alpha = \frac{R}{2L}$

$$\delta = \pi\frac{R}{\omega L}$$

The logarithmic decrement is thus equal to π times the ratio between the resistance and the inductive reactance.[7] It is also

---

[6] A few writers, Fleming in particular, define the logarithmic decrement as the natural logarithm of the ratio of two successive maxima in opposite directions. The Fleming decrement is thus equal to one-half the decrement as defined above.

[7] Reactance is ordinarily calculated from the simple expression for ω, which is not rigorously true when the decrement is very large.

equal to π times the ratio between the resistance and capacitive reactance,[1] for

$$\omega = \frac{1}{\sqrt{LC}} \text{ or } \omega L = \frac{1}{\omega C} \text{ and therefore}$$

$$\delta = \pi R \omega C.$$

Thus the logarithmic decrement is π times the reciprocal of the sharpness of resonance (discussed above on p. 36).

The logarithmic decrement expressed in terms of the three quantities resistance, inductance, and capacity is[2]

$$\delta = \pi R \sqrt{\frac{C}{L}}$$

Interpretations of Logarithmic Decrement.—The decrement has been given in terms of (1) a ratio of current amplitudes, (2) a ratio of impedance components, and (3) the reciprocal of the quantity called sharpness of resonance. A fourth interesting interpretation is in terms of an energy ratio. It is readily shown that $\delta = \frac{1}{2}$ the ratio of the average energy dissipated per cycle to the average magnetic energy at the current maxima, as follows:

$$\delta = \frac{\pi R}{\omega L} = \frac{R}{2fL} = \frac{\frac{RI^2}{f}}{2LI^2},$$

where I is the effective current as measured by an ammeter.

$\frac{RI^2}{f}$ = average energy dissipated per cycle, since RI^2 = average energy dissipated per second.

LI^2 = average (LI_k^2) (where I_k^2 = average current square during the k'th cycle), since I^2 = average I_k^2,

= average $\left(L\frac{I_{max}^2}{2} \right)$, (where I_{max} = maximum current during the k'th cycle), since $I_k^2 = \frac{1}{2}I_{max}^2$, just as in the case of undamped currents. But $\left(L\frac{I_{max}^2}{2} \right)$ is the magnetic energy at the current maximum during the k'th cycle. Therefore LI^2 = average magnetic energy at the current maxima. Hence the decrement is one-half the ratio of the average energy dissipated per cycle to the average magnetic energy at the current maxima. This is true when the energy is lost from the circuit by radiation as well as when lost by heating.

---

[1] Reactance is ordinarily calculated from the simple expression for ω, which is not rigorously true when the decrement is very large.

[2] This expression for the decrement does not hold when the decrement is extremely large.

Number of Oscillations in a Wave Train.—There are theoretically an infinite number of oscillations in a wave train. In practice, however, the wave train may be considered ended when the oscillations are reduced to a negligible amplitude. The fraction of the initial amplitude that is considered negligible depends on the use to which the oscillatory current is put. For a given ratio of initial amplitude to final amplitude, $\frac{I_0}{I_n}$, the number of complete oscillations n is given by

$$n = \frac{\log_e \frac{I_0}{I_n}}{\delta}$$

and the number of maxima or of semioscillations

$$= \frac{2 \log_e \frac{I_0}{I_n}}{\delta}$$

The oscillations after the amplitude is reduced to 0.01 of its initial value can usually be considered negligible. For $\frac{I_0}{I_n} = 100$, the number of oscillations $= \frac{4.6}{\delta}$. For example, the number of oscillations in which a current having $\delta = 0.2$ falls off to 1 per cent of its initial value is equal to 23.

26. PRINCIPLES OF DECREMENT MEASUREMENT

A number of so-called methods of measuring decrement are in reality measurements of resistance. From the resistance the logarithmic decrement of a circuit is calculated by $\delta = \frac{\pi R \sqrt{C}}{\sqrt{L}}$ or one of the related formulas. Any method for measuring the resistance of a circuit thus enables one to calculate the decrement. The value so obtained is the decrement of the current that would flow in the circuit if free oscillations were suddenly started, but not in general the decrement of the current used in making the measurement. The methods available for such resistance measurements are summarized below in sections 47 to 50. Only those methods in which damped oscillations are used can be considered actual measurements of decrement.

There are two classes of genuine decrement measurement, one in which free oscillations are used and one in which a damped electromotive force is impressed on the circuit so that both free and forced oscillations contribute to the current. Free oscillations are obtained in the case of pure impulse excitation. It is very difficult in practice to obtain such excitation. Assuming, however, that free oscillations are produced by the sudden discharge of the condenser at a constant potential difference N times per second, the effective current is given by equation (60),

$$I^2 = \frac{N}{4\alpha} I_o^2$$

Since

$$\alpha = \frac{\omega\delta}{2\pi} \text{ and } I_o^2 = \omega^2 C^2 V_o^2,\ I^2 = \frac{\pi N\omega C^2 V_o^2}{2\delta}.$$

This may be written, $I^2\delta =$ constant. Let I be the current for a certain resistance in the circuit, and I_1 be the current when a resistance R_1 is added so as to increase the decrement by an amount δ_1.

$$I^2\delta = I_1^2(\delta + \delta_1)$$
$$\therefore \delta = \delta_1 \frac{I_1^2}{I^2 - I_1^2} \qquad (61)$$

Bjerknes Methods.—A method of measuring the high-frequency resistance of a circuit using undamped currents has been described on page 38 above and others are described in sections 49 and 50

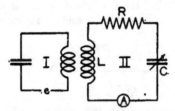

Fig. 69.—*Inductively coupled circuits for decrement measurements*

below. These methods have been extended to the measurement of resistance and decrement by the use of damped waves. When a damped emf is impressed on the circuit both free and forced oscillations exist, and the measurement is an actual measurement of decrement. For a circuit II (Fig. 69) very loosely coupled to a circuit I in which oscillations are generated, Bjerknes [*] showed that when the two circuits are in resonance,

$$I^2 = \frac{NE_o^2}{16L^2\alpha'\alpha(\alpha' + \alpha)}$$

* See reference No. 42, Appendix 2.

where E_0 is the maximum value of the impressed electromotive force, N the number of trains of waves per second, α' the damping factor of the emf due to circuit I impressed on circuit II, and α the damping factor of the second circuit. The circuit I may be a great distance from II, and may even be a distant radiating antenna. The equation holds only when α' and α are small in comparison with ω, or when the decrements δ' and δ are small compared with 2π.

In the resistance variation method of determining decrement, the resistance of the second circuit is increased by an amount R_1 changing α to $\alpha + \alpha_1$ and the original current I to some other value I_1. We then have

$$I_1^2 = \frac{NE_0^2}{16L^2\alpha'(\alpha+\alpha_1)(\alpha'+\alpha+\alpha_1)}$$

Since $\dfrac{\alpha'}{\alpha} = \dfrac{\delta'}{\delta}$, it follows that

$$\frac{I^2}{I_1^2} = \frac{(\delta+\delta_1)(\delta'+\delta+\delta_1)}{\delta(\delta'+\delta)} \tag{62}$$

This can be solved either for δ' if δ is known or for δ if δ' is known; δ' being the decrement of the wave emitted by circuit I and impressed on circuit II, δ being the decrement of circuit II, and δ_1 the amount by which δ is increased by adding resistance. The solution for δ' is

$$\delta' = \frac{2\delta\,\delta_1 + \delta_1^2 - \dfrac{I^2-I_1^2}{I_1^2}\,\delta^2}{\dfrac{I^2-I_1^2}{I_1^2}\,\delta - \delta_1} \tag{63}$$

This may be simplified by choosing the resistance inserted such that $\delta_1 = \delta$; then

$$\delta' = \delta\,\frac{4I_1^2 - I^2}{I^2 - 2I_1^2} \tag{64}$$

Another convenient simplified procedure is to vary the inserted resistance until the square of the current is reduced to one-half its previous value, then $\dfrac{I^2-I_1^2}{I_1^2} = 1$, and

$$\delta' = \frac{2\delta\,\delta_1 + \delta_1^2 - \delta^2}{\delta - \delta_1} \tag{65}$$

When δ is the unknown, the direct solution for δ in terms of δ' is complicated; equation (62) is in fact more convenient in this case than an explicit solution. A useful form of the solution for δ in certain cases is

$$\delta = \delta_1 \frac{KI_1^2}{I^2 - KI_1^2} \tag{66}$$

where $K = 1 + \dfrac{\delta_1}{\delta' + \delta}$

This is discussed on page 184 below. It should not be forgotten that these formulas apply only when the coupling is very loose and both decrements are small.

The reactance variation method of measuring decrement is similar in procedure to the resistance variation method, two observations of current being taken with different reactances. The method is described on pages 186 to 199 below, formulas (96) to (100) showing the method of calculation. The measurement of the decrement of a wave is treated in sections 54 and 55.

PART II.—INSTRUMENTS AND METHODS OF HIGH-FREQUENCY MEASUREMENT

❦

27. GENERAL PRINCIPLES

There is considerable difficulty in attaining high accuracy in measurements at radio frequencies. Much of this is due to the fact that the quantities to be measured or upon which the measurement depends are generally small and sometimes not definitely localized in the circuits. Thus the inductances and capacities used in the measuring circuits are so small that the effect upon these quantities of lead wires, indicating instruments, surroundings, etc., must be carefully considered. The capacity of the inductance coil and sometimes even the inductance within the condenser are of importance. In order to minimize these various effects it is generally best to use measuring circuits and methods which are the least complicated. On this account simple circuits and substitution methods in which the determination depends upon deflections are usually used in preference to more complicated methods.

In addition to the uncertainty or the distributed character of some of the quantities to be measured, there are other limitations upon the accuracy of radio measurements. The usual ones are the variation with frequency of current distribution, inductance, resistance, etc., and the difficulty of supplying high-frequency current of sufficient constancy. The latter limitation is entirely overcome by the use of the electron tube as a source of current but is troublesome when a buzzer, spark, or arc is used. As to the other difficulty, the variations of inductance, etc., with frequency, while these variations have a profound effect, they are generally subject to control. They must be reckoned with in the design of high-frequency ammeters and other instruments, but the quantities have definite values at a particular frequency under definite conditions, and their effect can usually be determined by calculation or measurement.

It is not always possible to determine the effects of the capacities of accessory apparatus and surroundings, nor to eliminate them,

96

and thus they remain the principal limitation upon the accuracy of measurements. These stray capacities include the capacities of leads, instrument cases, table tops, walls, and the observer. They may not only be indeterminate but may vary in an irregular manner. Means of dealing with them are discussed in section 31 below.

Relative Importance of Measurements.—On account of the requirement of simplicity in radio measurements, the methods available are quite different, and are fewer in number, than in the case of low-frequency or direct-current measurements. The wave meter and the ammeter are the principal measuring instruments used, and with these two most of the important radio measurements can be made. The principal additional pieces of apparatus are condensers and coils. These are the essential constituents of radio circuits. It may be recalled that the principal difference between the phenomena of high and low frequency is the importance of inductance and capacity at high frequency as compared with the predominance of resistance in low-frequency phenomena. Thus the resistance of circuits is the chief factor determining current flow in the case of direct current, and this is largely true in low-frequency alternating current; in high-frequency circuits, on the contrary, resistance is of minor importance, and the flow of current is mainly determined by reactance, a quantity dependent upon inductance and capacity. The significance of the rôle played by reactance is shown by sections 13–18 above, in which many problems of radio circuits are worked out by the use of reactance without any consideration of resistance.

The increase of resistance with frequency renders all the more striking the general fact that resistance is of less importance in determining the distribution of high-frequency currents than capacity and inductance. While resistance is not of primary, importance in radio circuits, nevertheless it must in many cases be known accurately, and on account of the change with frequency must be measured at the particular frequency for which its value is needed.

WAVE METERS

28. THE FUNDAMENTAL RADIO INSTRUMENT

The most generally useful measuring instrument at radio frequencies is the wave meter. It measures primarily frequency, which is customarily expressed in terms of wave length. It may also be used, however, to compare inductances and capacities, to measure resistance and decrement, to determine resonance curves, and, in fact, to make most of the measurements required for radio work. When equipped with a buzzer or other source of power, a wave meter may also be used as a generator of currents of known frequency. Wave meters may in general be divided into two classes, standard and commercial. The former type is so designed as to permit its calibration to be derived from a knowledge of the value of its constituent elements; the latter is designed for compactness and convenience, and is ordinarily calibrated by comparison with a standard.

In its usual form a wave meter is essentially a simple radio circuit, consisting of an inductance coil and condenser in series, with an ammeter or other device to indicate either the current flowing in the circuit or the voltage across the whole or a part of the capacity or inductance. Either the inductance or capacity is made variable and sometimes both. Usually the condenser is variable, and a number of inductance coils of different values are provided. Each coil in connection with the condenser gives a definite range of wave lengths, and the different coils are so chosen that these separate ranges overlap so as to cover the complete range of wave lengths desired.

To measure the wave length of the oscillations given out by a source the wave meter is loosely coupled to it and the variable condenser adjusted to that setting which gives the maximum current in the indicating device. In this condition of resonance the frequency is given by equation (24), page 32, and the wave length is related to the inductance and capacity in the circuit according to the expression

$$\lambda = k\sqrt{LC} \tag{67}$$

in which k is a constant, values of which are given on page 312 for L and C in various units. In a standard wave meter the values L and C are known and the wave length may be computed. In the commercial type the values of the wave length are determined for each coil and each condenser setting by comparison with a standard wave meter, and may be read off from curves or from the scale of the variable condenser. This scale may be graduated in degrees, arbitrary divisions, wave length, or even in terms of frequency or of capacity. Sometimes several scales are put on one instrument.

29. CALIBRATION OF A STANDARD WAVE METER

The most direct method for the wave length calibration of a standard or commercial wave meter is a comparison with a high-frequency alternator. From the speed of the machine and the number of poles or other structural data the frequency of alternation can be computed directly. The range of such alternators is, however, limited; the usual construction does not furnish a wave length shorter than 3000 meters.

Low-Frequency Measurements as Basis.—Another method of procedure consists in measuring separately the inductance and capacity of the standard wave meter and computing the wave length from these quantities when combined. The capacity is measured at low frequencies and assumed to be the same at high frequencies. This assumption is justifiable if the condenser is one that is free from dielectric absorption and has an insulation resistance so high as not to affect the low-frequency capacity. A convenient means of measuring capacity with low-frequency interrupted direct current is the Maxwell bridge. This method [10] permits the determination of capacity in terms of resistance and the frequency of the commutator, tuning fork, or other device which charges and discharges the condenser. Capacities can be determined by this method to an accuracy of 1 micromicrofarad, which is sufficient for radio measurements.

The capacity of an unshielded condenser will depend upon its position in the circuit and its proximity to the body of the observer, walls of the room, etc. The condenser should therefore have one set of its plates connected to a metal shield, and the shield connected to ground, both when it is being calibrated at low frequency and when it is being used in the wave-meter circuit. The ground connection to a radio circuit should be a thick conductor, run by as short a path as possible to ground.

The inductance coils, if wound with properly stranded wire so that there is little skin effect and if free from other metal so that there is no eddy-current effect, may have their inductances computed or measured at low frequency and the values used at high frequencies; provided, however, that condensers of such large capacity values are used with the coils that the capacities of the coils themselves are negligible. In certain coils of small radius and wound with heavy stranded conductor with inferior insulation

---

[10] See reference No. 174, Appendix 2.

between strands it was found that at radio frequencies the inductance was reduced by more than 1 per cent.

Use of Calculable Inductance Standard.—A good method of determining the high-frequency inductances of the standard coils is to compare at high frequencies the smallest coil of the wavemeter with an inductance of simple form such as a large single-turn square or rectangle, the value of which can be computed. (Use formulas in sec. 68.) The larger coils are then compared with the small coil as outlined below. If the square is made of round wire the inductance may be calculated very accurately by (137). The following corrections must be considered. (See Fig. 70.)

First, the condenser calibration takes account of the capacity only from the terminals A, B, the binding posts of the condenser.

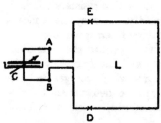

FIG. 70.—*Circuit consisting of a calculable inductance standard and a shielded standard condenser*

In addition to this there is the capacity between the leads to the square and between the leads, square, etc., and the shield or case of the condenser. This correction may be determined experimentally by two methods. In one the square is cut, say at the points D and E, and some other coil connected to the terminals A, B, and coupled to a source of oscillations. Keeping the wave length of the source constant the setting of the condenser C for resonance is obtained with the terminals of the square connected to the condenser terminals and then with them disconnected. The capacity of the condenser for resonance in the latter case will be greater than in the former by an amount very close to the required correction.

Capacity Correction Determined by Pliotron Harmonics.—In the other method the harmonics emitted by such a source as the pliotron generator are used. In the current furnished by such a generator all of the harmonics are present. The circuit is first tuned to the fundamental by means of the condenser C and then to the harmonic, say of double frequency. Since the inductance is kept constant and the frequency varies inversely as the square root of the capacity, the total capacity required to tune for the harmonic will be one-quarter of that for the fundamental. The method depends on the assumption that the frequencies of the harmonics are exact integral multiples of the fundamental frequency. Letting C_f and C_{2f} represent the capacities of the condenser for the fundamental and harmonic, and c the extraneous capacity, we have

$$(C_f + c) = 4(C_{2f} + c)$$

therefore

$$c = \frac{C_f - 4C_{2f}}{3}$$

Determination of Inductance Corrections.—The second correction which must be considered involves the inductance of the leads connecting the square and the condenser, the inductance of the leads within the condenser, and the inductance of the indicating instrument. The first of these inductances can be calculated with sufficient accuracy, making use of the formulas for straight wires and two parallel wires given on page 245. The inductance of the leads within the condenser is very small in a well-designed condenser. It can be determined by comparing two squares of different inductances, one of which should be small. If L_1 and L_2 are the inductances of the squares and leads, l is the inductance of the condenser and C_1 and C_2 are the values of capacity at which resonance is obtained with the two squares at a given wave length, then

$$(L_1 + l)C_1 = (L_2 + l)C_2$$

and

$$l = \frac{L_1 C_1 - L_2 C_2}{C_2 - C_1}.$$

If the indicating instrument is inserted directly in the circuit, its inductance, if appreciable, must be added to that of the square. A sensitive hot-wire ammeter may have 0.5 microhenry inductance, which may be a large fraction of the inductance of the circuit. The value can be readily determined by inserting the ammeter in a circuit previously tuned to resonance by means of another indicating instrument and noting the change in the condenser setting required to retune the circuit. A crossed-wire thermoelement, if made with very short heating wires and leads, has a very small inductance and may usually be inserted directly in the circuit without appreciable error. Instead of inserting the indicating instrument directly in the circuit, it is possible to connect one or more turns of wire to the instrument and couple it with the circuit. If sufficient power is available, the coupling may be made so loose that the indicating instrument has no appreciable effect upon the inductance of the circuit.

Comparison of Inductance Standards.—The method of comparison of the standard square and a standard coil is shown in Fig. 71. *I* is the source, *II* the circuit of the square and condenser, and *III* the standard coil and condenser. The source *I* is set at a given wave length and the condenser of *II* is tuned to resonance, and the condenser reading obtained. Then *II* is detuned and *III* is tuned and the condenser read. Both the inductance and capacity are known for the circuit *II*, and since *III* is tuned to the same wave length the product of capacity and inductance for this circuit must

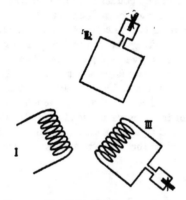

Fig. 71.—*Circuits involved in the comparison of inductance standards*

be the same as that for *II*. Dividing this product by the observed capacity for *III* gives the apparent inductance in *III*. This process is repeated over a wide range of wave lengths and the apparent inductance of the coil is found to increase as the wave length is reduced. It is shown above (p. 64) that this is a result of the fact that the capacity of the coil has been included, together with its inductance; hence the term ''apparent'' inductance is used. It is shown in section 19 that, instead of using the values of the apparent inductance for each wave length, it is possible to represent the coil very accurately by a fixed value for the inductance of the coil called the ''pure'' inductance and a value for the coil capacity. When the coil is used with a condenser to form a wave-meter circuit, the capacity of the coil is added to that of the condenser and the wave length computed from this total capacity and the pure inductance. In the absence of skin effect, etc., the pure inductance of the coil is identical with its low-frequency inductance. Having determined the pure inductance and capacity of the smallest coil of the standard wave meter, this coil may be used to determine the values of the other coils by the same comparison method. The small coil with a large condenser is tuned to the wave length of the source, and a larger coil with smaller condenser is then tuned to the same wave length. The product $L_1 C_1$ for the small coil, which is known, is equal to the product $L_2 C_2$ for the larger coil from which the apparent inductance of the larger coil

$$L_2 = \frac{L_1 C_1}{C_2}.$$

The apparent inductance of all the larger coils may thus be determined by stepping from coil to coil.

Coil Measurements by Pliotron Harmonics.—Some errors will arise in these successive comparisons which makes it desirable to be able to compare directly the large coils with the small. By making use of the harmonics of a generator of the pliotron type it is possible to compare directly small and large coils without the requirement of very large condensers, and the capacity of each coil can be readily determined. In comparing the coils, a circuit containing the large coil is tuned to the fundamental. The circuit with the small coil is then tuned to the harmonic of frequency two, three, or more times the fundamental. The wave length in the latter case will be one-half, one-third, etc., of that in the former; the product of L and C will be one-fourth, one-ninth, etc. Thus, if C_2 is the known capacity, L_2 the unknown inductance for the large coil, and L_1 and C_1 are the known values for the small coil and the harmonic of threefold frequency is used, we have

$$L_2 C_2 = 9 L_1 C_1$$

$$L_2 = 9 \frac{L_1 C_1}{C_2}.$$

L_2 is the apparent inductance. The capacity of the coil is then obtained in a manner similar to that given above as one of the methods of getting the capacity of leads, etc., of the square. The coil plus condenser is tuned to the fundamental and then to the harmonic of double frequency. The condenser capacity will be

roughly one-quarter in the latter case and should be quite small.
As above, we obtain for the coil capacity $c = \dfrac{C_f - 4C_{2f}}{3}$. It is
advisable to repeat the measurement at several differing fundamental frequencies and average the values of c obtained. The
pure inductance L_p is then determined, making use of known
values of L_2C_2, as determined above, for

$$L_p(C_2 + c) = L_2C_2$$

$$L_p = L_2\frac{C_2}{C_2 + c}.$$

The mean based on several values should be obtained.

In the above comparisons the leads of the coils have been considered as a part of the coil, contributing both to the inductance
and capacity of the whole. These leads should be fixed and
definite and should be of sufficient length and otherwise designed
so that the coil constants will not be appreciably altered on
account of eddy currents in the condenser case or capacity to it,
when the coil is connected to the condenser to form a wave meter.

30. STANDARDIZATION OF A COMMERCIAL WAVE METER

A commercial wave meter is generally equipped with one or
more indicating devices, of which the hot-wire ammeter and
crystal detector with phones are especially important. In
addition, it is customary to provide a buzzer circuit, so that
oscillations of known wave length may be generated by the
wave meter. The indications of the wave meter will be somewhat different, depending upon the way it is operated—that is,
whether one or the other of the indicating devices is used or
whether it is used as a source of oscillations. Hence, in each of
these cases it will usually be necessary to have a separate calibration. Calibration of wave meter used as a source is treated on
page 108 below.

Wave Meter with Ammeter.—When the hot-wire ammeter is in
use, it is either inserted in the circuit directly, when, on account
of its high resistance, it is generally shunted by a small inductance,
or it may be tapped across a number of turns of the inductance
coil. The calibration is effected by comparing the wave meter
with the standard wave meter in a manner similar to that outlined
above for intercomparing the standard wave-meter circuits. The
condenser of the commercial instrument is set at a given reading

and the wave length of the source adjusted until the ammeter of the wave meter indicates maximum current. The wave meter is then detuned, and without changing the source the standard wave meter is adjusted until resonance is obtained. The wave length as indicated by the standard corresponds to the chosen setting of the commercial wave meter, and, repeating the observations for other settings, a curve may be obtained giving the wave length as a function of the setting. Or if it is desired to engrave the scale of the commercial instrument so as to read wave lengths directly, the standard circuit may be set at a chosen integral wave length, the source adjusted to this wave length, and the corresponding setting of the commercial instrument found and

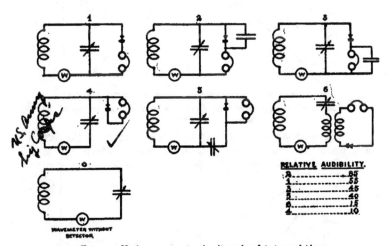

FIG. 72.—*Various wave-meter circuits, using detector and phones*

marked. In these comparisons a source of either damped or undamped oscillations may be used. The latter, however, permits a higher precision in the measurement on account of the sharper tuning. The ammeter of the commercial wave meter can only be used when it is possible to draw a considerable amount of power from the source. It is practically indispensable when the wave meter is to be used for measurements of resistance or logarithmic decrement.

Use of Crystal Detector.—When the source supplies only a small amount of power, it is necessary to use a sensitive indicator, such as a crystal detector and phones. When such a detecting circuit is connected or coupled to the wave-meter circuit, the wave

length calibration and the resistance of the wave meter will be
changed somewhat, depending upon the type of detecting circuit.
The changes will also depend to some extent upon the adjust-

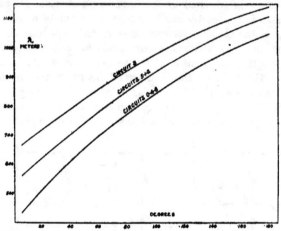

FIG. 73.—*Increase of wave length for different condenser settings due to the addition of detector circuit*

ment of the crystal contact, so that it is important in the design
of a wave meter to choose a detecting circuit which will least

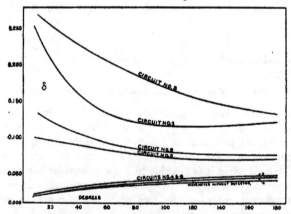

FIG. 74.—*Increase in decrement of the wave-meter circuit due to the detector circuit*

affect the wave-meter constants. The wave meter should then
be calibrated with the detecting circuit connected as for use.
Circuit o in Fig. 72 represents the wave meter without detector,

while the circuits numbered 1–6 show the detecting circuits frequently used.

Typical examples of the effects of these circuits are shown in Figs. 73 and 74. Fig. 73 illustrates the increased wave length for different condenser settings caused by the addition of the detector circuit. The increase is extremely small in the case of circuits 4 and 6—i. e., when the detecting circuit is connected to the wave-meter circuit at one point or when it is loosely coupled

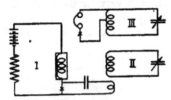

FIG. 75.—*Arrangement of circuits for comparing wave meters by impact excitation*

to the wave-meter circuit. In Fig. 74 the effect of the detector circuit in increasing the decrement (or resistance), and hence in impairing the sharpness of tuning of the wave meter is shown, and here again the circuits 4 and 6 appear to produce the least effect. As shown by the relative audibility values in Fig. 72, these circuits are not as sensitive as those which withdraw more power from the main circuit.

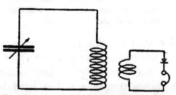

FIG. 76.—*Arrangement of circuits for determining resonance by coupling to an aperiodic detector circuit*

Use of Buzzer.—In order that an audible note may be heard in the phone, the calibration must be carried out using either damped oscillations with a wave-train frequency that is audible or undamped oscillations that are interrupted or "chopped" at an audible frequency. A simple and accurate method of comparison is that shown in Fig. 75. Here circuit *I* is the buzzer circuit described on page 227, which excites the standard wave-meter circuit *II* by impact excitation. The commercial wave meter *III* is loosely coupled to the standard, and resonance is indicated by the setting for maximum response in the phones. Sharpness in setting is facilitated by reducing the coupling between *II* and *III* until the phone responds only when *III* is very nearly in resonance with *II*. The standard circuit has no detector or buzzer attached; hence, its calibration is unaffected if it is sufficiently loosely coupled to *I* and *III*. The buzzer circuit is generally not strictly aperiodic and will show a very broad tuning at its natural frequency. It is necessary to use it at frequencies differing considerably from its natural frequency or errors will be introduced.

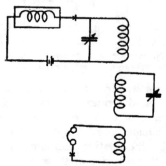

FIG. 77.—*Arrangement of circuits for comparing a wave meter with a standard buzzer circuit, the resonance point being indicated by the aperiodic detector circuit*

Another method makes use of a tuned buzzer source, and each circuit is separately tuned to the source in the same manner as described above for the comparison when ammeters are used. Resonance in the standard circuit is indicated by an aperiodic detector circuit loosely coupled to that circuit as indicated in Fig. 76. When the commercial wave meter is used as a source, the buzzer is usually connected as shown in Fig. 77, circuit *I*. The leads to the buzzer will add capacity to the circuit and the lengths of the waves emitted will be increased, in particular at low condenser settings. The calibration is simply carried out by the circuits of Fig. 77. Circuit *II* is the standard wave meter and *III* an aperiodic detector circuit loosely coupled to the standard circuit. The coil of circuit *III* may be so oriented as not to be directly affected by circuit *I*. Then either circuit *II* may be set at integral wave lengths and the settings found at which these wave lengths are emitted by I, or I is set and the wave length corresponding to the setting is found by tuning II.

CONDENSERS

31. GENERAL

A condenser is an apparatus so designed that electrostatic capacity is its important property. It consists of a pair of conductors with their surfaces relatively close together, separated by an insulating medium called the dielectric. When the two conducting plates are parallel, close together, and of large area, the capacity of a condenser is given by

$$C = 0.0885 \times 10^{-8} \frac{KS}{\tau},$$

where C is in microfarads, S = area of one side of one conducting plate in cm^2, τ = thickness of dielectric between the plates in centimeters, and K, the dielectric constant, = 1 for air and is between 1 and 10 for most ordinary substances. Formulas for capacities of various combinations of conductors, antennas, etc., are given on pages 237 to 242. These formulas assume that the charge is uniformly distributed over the surfaces of the conductors, no corrections being made for edges or end effects. It is seldom worth while, however, to apply a correction on this account, because the capacity to the condenser case or other conductors is ordinarily not calculable, so that the actual capacity of a condenser can be calculated only approximately. The actual value is likely to be in excess of that calculated. When very accurate values are required they must be obtained by measurement. The usual methods for measuring capacities at radio frequencies are discussed on pages 129 to 131.

Series and Parallel Connection.—When two or more condensers are connected in series, the resultant capacity is given by

$$\frac{1}{C} = \frac{1}{C_1} + \frac{1}{C_2} + \frac{1}{C_3} + \quad \cdot \quad \cdot \quad \cdot$$

The resultant capacity of a number of condensers connected in series is always less than the smallest capacity in the series. The series connection is used when it is necessary to use a voltage higher than a single condenser would stand without breakdown.

When condensers are connected in parallel, their capacities are simply added, thus:

$$C = C_1 + C_2 + C_3 + \quad \cdot \quad \cdot \quad \cdot$$

The laws of series and parallel combination of condensers are thus the inverse of the laws for resistances. In combining condensers, care must be exercised that there are no appreciable mutual capacities between the parts combined.

Stray Capacities.—It is very difficult to concentrate the total capacity in a radio circuit at a particular point in the circuit. Every part of the apparatus has capacities to other parts, and these small stray capacities may have to be taken into account as well as the capacity of the condenser which is intentionally inserted in the circuit. The stray capacities are particularly objectionable because they vary when parts of the circuit or conductors near by are moved. Thus, they make it difficult to keep the capacity of the circuit constant. The disturbing effects may be minimized in practice, as follows: (1) Keeping the condenser a considerable distance away from conducting or dielectric masses; (2) shielding the condenser, i. e., surrounding the whole condenser by a metal covering connected to one plate; (3) using a condenser of sufficiently large capacity so that the stray capacities are negligible in comparison. The first of these methods reduces only the stray capacities of the condenser itself to other parts of the circuit. This is also true of the second method, which is none the less a desirable precaution. One of the chief causes of variation in the stray capacities is the presence of the hand or body of the operator near some point of the circuit. Shielding the condenser reduces the capacity variation from this cause. The third method is, in general, the best for reducing or eliminating these errors. On account of the stray capacities of its various parts, the whole circuit is in effect a part of the condenser, and their effect is best rendered negligible by making the condenser capacity relatively great.

Imperfection of Condensers.—In an ideal condenser, the conductors or plates would have zero resistance and the dielectric infinite resistivity in all its parts. In case an alternating emf is applied to such a condenser, current will flow into the condenser as the voltage is increasing and flow out as the voltage is decreasing. At the moment when the emf is a maximum no current will be flowing, and when the emf is zero the current will be a maximum. Hence, in a perfect condenser the current and voltage are 90° out of phase. In actual condensers the conditions as to resistance in the plates and dielectric are not fulfilled, and in consequence an alternating current flowing in a condenser is not exactly 90° out of phase with the impressed voltage. The difference between 90° and the actual phase angle is called the "phase difference." In an ideal condenser there would be no consumption of power; the existence of a phase difference means a power loss, which appears as a production of heat in the condenser. The amount of the power loss is given, as for any part of a circuit, by $P = EI \cos \theta$, where θ is the phase angle between current and voltage and $\cos \theta$ is the power factor. This is equivalent to

$$P = EI \sin \psi,$$

where ψ is the phase difference and $\sin \psi$ is the power factor of the condenser. In all except extremely poor condensers, ψ is small, $\sin \psi = \psi$, and thus the phase difference and power factor are synonymous. The power loss is given by

$$P = \omega C E^2 \sin \psi \tag{68}$$

This shows that, for constant voltage, the power loss is proportional to the frequency, to the capacity, and to the power factor. Information on the power factors of condensers is given in section (34) below.

Change of Capacity with Frequency.—Another effect of the imperfection of dielectrics is a change of capacity with frequency. The quantity of electricity which flows into a condenser during any finite charging period is greater than would flow in during an infinitely short charging period. In consequence the measured or apparent capacity with alternating current of any finite frequency is greater than the capacity on infinite frequency. The latter is called the geometric capacity (being the capacity that would be calculated from the geometric dimensions of the condenser on the assumption of perfect dielectric). The capacity of a condenser decreases as the frequency is increased, approaching the geometric capacity at extremely high frequencies. For this reason, when dielectric constants are measured at high frequencies of charge and discharge, smaller values are obtained than with low frequencies.

When the phase difference of the condenser is due to ordinary leakage or conduction through the dielectric or along its surface, the apparent capacity at any frequency is readily shown to be

$$C = C_0 (1 + \psi^2) = C_0 \left(1 + \frac{10^{12}}{R^2 \omega^2 C_0^2}\right)$$

where C_0 is the geometric capacity in microfarads, ψ the phase difference, and R the leakage resistance in ohms. It is evident that the apparent capacity decreases very rapidly as frequency increases. For example, suppose a condenser whose geometric capacity is 0.001 microfarad to have a leakage resistance of 10 megohms, the dielectric being otherwise perfect. Its capacity at 60 cycles will be 0.001070 microfarad, at 300 cycles will be 0.001003 microfarad, and at all radio frequencies will be equal to the geometric capacity.

When the phase difference of a condenser is due to dielectric absorption (a phenomenon discussed below, p. 124), the capacity decreases as the frequency increases, as before, approaching the geometric capacity at infinite frequency, but the amount of the change can not be predicted from a knowledge of the phase difference. The change with frequency is large in condensers that have large phase difference. In certain cases the change of capacity with frequency has been found to be roughly proportional to the reciprocal of the square root of frequency.

A series resistance in the plates or leads of a condenser causes a phase difference but does not give rise to a change of capacity with frequency.

When the leads inside the case of a condenser are long enough to have appreciable inductance the capacity measured at the terminals appears to be greater than it actually is. The magnitude of the effect is given by

$$C_a = C[1 + \omega^2 CL(10)^{-12}]$$

where C_a is the apparent or measured capacity, and in the parenthesis C is in microfarads and L in microhenries. Thus, the inductance of the interior leads makes the apparent capacity of a condenser increase as frequency increases, while the imperfection of the dielectric makes the capacity decrease with increase of frequency.

32. AIR CONDENSERS

Electrical condensers are classified according to their dielectrics. The plates are relatively unimportant, their only requirement being low resistance. This requirement is met in the materials used for condenser plates, viz, aluminum, copper, brass. When the plates are thin the material must not have too high a resistivity. Various dielectrics are used; the one most frequently used in radio measurements is air.

A condenser which is to be used as a standard of capacity in measurements at radio frequencies is itself standardized at low frequencies, and its construction must be such that either the capacity does not change with frequency or the change can be calculated. The capacity of a condenser with a solid dielectric changes with the frequency in an indeterminate manner, and hence it is practically impossible to calculate the capacity at high frequencies from that measured at low frequency. Air is very nearly a perfect dielectric, hence a condenser with only air as a dielectric should show no change in capacity with the frequency, and thus the capacity at radio frequencies should be the same as that for low frequency. It is on this account that air condensers are quite generally used as standards of capacity in radio measurements.

Phase Difference of Air Condensers.—Air condensers are valuable in radio measurements for another reason. Their perfection, from a dielectric standpoint, involves freedom from power loss. The phase difference of an ideal air condenser is zero; there is no component of current in phase with the electromotive force, and thus the condenser acts as a pure capacity and introduces no resistance into the circuit. It is consequently advantageous to use them in circuits in which it is desirable to keep the resistance very low. In resistance measurements at high frequencies it is often necessary to assume that the resistance of the condenser in the circuit is negligible. This requires the use of a properly constructed air condenser.

Only the most careful design, however, can produce an air condenser which is close to perfection. In order to support the two conductors or sets of plates and insulate them from each other it is necessary to introduce some solid dielectric. There is necessarily some capacity through this dielectric, and since all solid insulators are imperfect dielectrics this introduces a phase difference. The magnitude of the phase difference is determined by the quality of the solid dielectric and the relative capacities through this dielectric and through the air. The effect is magnified by the concentration of the lines of electric field intensity in solid dielectric due to its high dielectric constant. Some air condensers tested by this Bureau, in which the pieces of dielectric used as insulators were large and poorly located, had phase differences or power factors many times greater than those of commercial paper condensers. Although purporting to be air condensers they were actually poorer than ordinary solid-dielectric condensers because the insulating pieces used to separate the plates were very poor dielectrics. In a variable air condenser the phase difference varies with the setting and is approximately inversely proportional to the capacity at any setting. The equivalent resistance (defined on p. 125) is inversely proportional to the square of capacity at any setting.

The questions of materials and construction of air condensers are further dealt with below in connection with design. It is seldom safe to assume that an air condenser has zero phase difference.

Antennas are subject to this same imperfection. An antenna is essentially an air condenser and is similarly subject to power loss from poor dielectrics [11] in its field.

Simple Variable Condenser.—In the most generally used types of air condensers the capacity is continuously variable. Variable condensers are extensively used, because most radio measurements involve a variation of either inductance or capacity, and it is relatively difficult to secure sufficient variation of an inductance without variation of the resistance and capacity in the circuit. The most familiar type of variable air condenser has two sets of semicircular plates (see Fig. 78, facing p. 118), one set of which can be revolved, bringing the plates in or out from between the plates of the fixed set. The position of the movable plates is indicated by a pointer moving over a scale which is marked off in arbitrary divisions. These may be degrees, 0° corresponding to the position of the plates when they are completely outside of the fixed set and the capacity is a minimum, and 180° when the capacity is a maximum. It is preferable to divide the range into 100 divisions rather than into degrees. The capacity of a condenser is proportional to the area of the plates. In a variable condenser of the semicircular type the effective area of the plates is changed by rotating the movable plates and, neglecting the edge effects, it is changed in proportion to the angle of rotation. As a result the capacity is approximately proportional to the setting throughout a wide range, provided that the condenser is well constructed and the distance between the two sets of plates is not affected by rotation of the movable set. Fig. 79 shows a typical capacity curve for such a condenser. Throughout the range in which the capacity curve is a straight line the capacity is given by the formula

$$C = a\theta + b$$

---

[11] See reference No. 197, Appendix 2.

where a and b are constants and θ is the setting in scale divisions. The coefficient a represents the change in capacity for one division and may be computed by taking the difference of the capacities at, say, 20 and 80 and dividing by 60. The constant b represents the capacity which the condenser would have at 0 if its linear character were maintained down to this setting. It may be positive, negative, or zero, depending upon the setting of the pointer relative to the movable plates. Its value is determined by subtracting the value of $a\theta$ at, say, 30 from the actual value of the capacity at that setting. When a condenser has a capacity curve that is closely linear, it may be found easier to compute

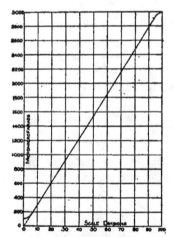

Fig. 79.—*Typical capacity curve for condenser with semicircular plates*

the capacity for given settings by means of the formula rather than to read off the values from a curve.

Uniform Wave-Length Type.—Some condensers have been specially designed to give a capacity curve different from the customary linear curve. In a wave meter in which a semicircular plate condenser is used, for any one coil the wave length varies as the square root of the capacity and hence as the square root of the setting of the condenser. If it is attempted to make the wave meter direct reading—i. e., to substitute a wave-length scale in place of the scale of setting in degrees—this wave-length scale will be nonuniform and either crowded together too closely at the low settings or too open at the high settings. In order to obtain a uniform scale of wave lengths, it is necessary to have the capacity vary as the square of the displacement or rotation. Tissot[12] proposed a condenser which had two sets of square plates (Fig. 80) which moved relative to each other along their diagonals. The same result can be obtained in a rotary condenser if one set of plates is given the proper shape.[13] It is required that the capacity and, hence, the effective area between the plates shall vary as the square of the angle of rotation. Thus,

FIG. 80.—*Form of condenser plates for which the capacity varies as the square of the displacement*

$$C = A = a\theta^2$$

But in polar coordinates the area is equal to

$$A = \frac{1}{2}\int r^2 d\theta$$

Differentiating these two values of A,

$$\frac{dA}{d\theta} = \frac{1}{2}r^2 = 2a\theta,$$

$$r = \sqrt{4a\theta}$$

In the condenser as actually made the fixed plates may be semicircular and the moving plates given the required shape to

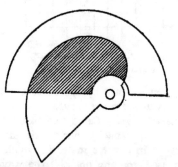

FIG. 81.—*Form of rotary condenser plates for which the capacity varies as the square of angular displacement*

make the effective area vary as the square of the angle of rotation. This effective area is the projection of the moving plates on the fixed. To provide clearance for the shaft of the moving-plate system a circular area of radius r_2 must be cut from the fixed

---

[12] See reference No. 81, Appendix 2.   [13] See reference No. 82, Appendix 2.

plates, and, taking this into account, the equation of the boundary curve of the moving plates becomes

$$r = \sqrt{4a\theta + r_3^2}$$

In Fig. 81 the form of the plates is shown and the effective area denoted by shading.

Decremeter Type.—Another special shape of plates is utilized in the direct-reading decremeter [14] developed at this Bureau. As shown in section 55, logarithmic decrement may be measured by the per cent change in capacity required to reduce by a certain amount the indication of an instrument in the circuit at resonance. In order that equal angular rotations may correspond to the same decrement at any setting of the condenser, it is necessary that the per cent change in capacity for a given rotation shall be the same at all parts of the scale. Thus we have the requirement

$$\frac{dC}{C} = ad\theta;\quad a = \text{constant} = \text{per cent change of capacity per scale}$$

division. By integration,

$$\log C = a\theta + b \text{ where } b = \text{a constant, or}$$
$$C = \epsilon^{(a\theta + b)} = C_0 \epsilon^{a\theta}$$

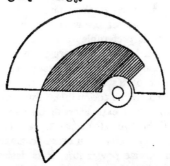

Fig. 82.—*Form of rotary condenser plates in which the per cent change of capacity is the same throughout the entire range of the condenser*

$C_0 = \epsilon^b = $ capacity when $\theta = 0$.
Since the area must vary as the capacity

$$A = \frac{1}{2}\int r^2 d\theta = C_0 \epsilon^{a\theta}$$
$$\frac{dA}{d\theta} = \frac{r^2}{2} = C_0 a \epsilon^{a\theta}$$
$$r = \sqrt{2 C_0 a \epsilon^{a\theta}} \tag{69}$$

[14] See reference No. 196, Appendix 2, and description of decremeter on p. 199.

This latter is then the polar equation of the bounding curve required to give a uniform decrement scale.

The shape of the condenser, as actually made, is shown in Fig. 82 and Fig. 218, facing page 320. A small semicircular area is omitted from the fixed plates, to provide clearance for the metal washers which must hold the moving plates together. Taking account of this omitted area, of radius r_2, the equation of the boundary curve becomes

$$r = \sqrt{2C_0 a e^{a\theta} + r_2^2}$$

The effective area is denoted by shading in Fig. 82.

A wave-length scale placed on such a condenser is somewhat crowded at high settings—just the opposite of the effect with a semicircular plate condenser. It is much more nearly uniform than in the case of a semicircular-plate condenser, as might be expected from the similarity of shape of Figs. 81 and 82.

Important Points in Design.—In a standard condenser it is required that the capacity remain constant and be definite. The former condition requires rigidity of construction, which is difficult to secure in a variable condenser. The pointer and movable plates must be securely fastened to the shaft so that no relative motion is possible. A simple set screw is not sufficient to hold the pointer in place. It is preferable to have no stops against which the pointer may hit. Particular care must be exercised in insulating the fixed and moving plates from each other. The suspension of heavy sets of plates from a material such as hard rubber, which may warp, is objectionable. In some cases the high temperature coefficient of expansion of insulators may produce relative motions of the two sets of plates, resulting in a high temperature coefficient of capacity. In order to make the capacity definite and also minimize power loss, it is desirable to surround the condenser with a metal case which is connected to one set of plates and grounded when the condenser is calibrated and used. The inductance of the leads inside of the condenser should be a minimum, for the apparent capacity of a condenser at high frequencies will increase with the frequency due to inductance in the leads in a similar manner to the variation of apparent inductance of a coil with distributed capacity. The connections from the binding posts to the plates should therefore be short and thick; this minimizes both inductance and resistance.

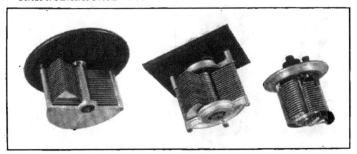

FIG. 78.—*Commercial types of variable air condensers*

FIG. 84.—*Quartz-pillar standard condenser of fixed value*

FIG. 85.—*Type of variable condenser suitable for high voltages, with double set of semicircular moving plates*

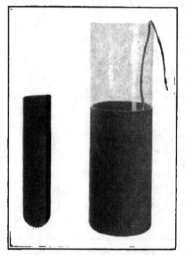

FIG. 86.—*Leyden jar type of high voltage condenser*

FIG. 87.—*Mica condensers suitable for high voltages*

The resistance of leads and plates and contact resistances between the individual plates and separating washers should be kept as low as possible, in order to minimize the phase angle. For the same reason the dielectric used to support one set of plates and insulate the two sets from each other should be as nearly perfect as possible. The insulation resistance must be high, because otherwise it would introduce an error in the measured capacity at low frequencies (see p. 111). Power loss in the

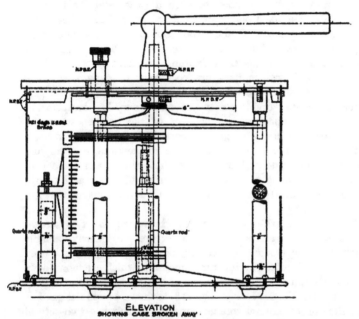

ELEVATION
SHOWING CASE BROKEN AWAY

FIG. 83.—*Quartz-pillar variable condenser used as primary standard of capacity at high frequencies*

dielectric should be kept small by using suitable insulating material, locating it where the electric field intensity is not great, and using pieces of such size and shape that the capacity through it is small. Built-up mica, bakelite, formica, and similar materials have been found to have bad dielectric losses. Hard rubber and porcelain are better. For further information on dielectrics, see section 34 below, on Power Factor. In special cases quartz may be used, as in the special condensers next described.

Bureau of Standards Type.—In the customary design of variable air condensers, the movable plates are insulated from the fixed by rings of insulating material around the bearings at top and bottom. This introduces some difficulties in the design of the bearing; the capacity through the dielectric is likely to be large and the choice of insulator is limited.

In the condensers used as standards at this Bureau, instead of the moving plates being insulated from the case of the condenser, they are in electrical connection with the case and the bearings are metal. The fixed plates are insulated from the moving plates and the case by pieces of quartz rod inserted in the uprights which support the fixed plates. The capacity through the insulator is small. The use of quartz is desirable on account of very high insulation and low temperature coefficient of expansion. The condensers of this type which have been tested have shown permanence, no measurable power loss, and a very low temperature coefficient of capacity. On account of the brittleness of the quartz and the heavy weight supported by it, these condensers are not very portable and must be dismounted in shipping. Views of the inside and the case of these condensers are shown in Fig. 214, facing page 318. Two sizes, of 0.007 and 0.0035 microfarad capacity, are shown. Data on the construction are given in Fig. 83.

Fixed Condensers.—Air condensers in which the two sets of plates are fixed relative to each other are particularly valuable when well constructed on account of their permanence. They serve to standardize variable air condensers, which are more likely to change, and are of value in standardizing solid-dielectric condensers. In Fig. 84, facing page 118, is shown a type designed at this Bureau, of 0.01 microfarad capacity, which has two sets of square plates and which is insulated by quartz supports in a manner similar to the variable condensers just described. Great care is necessary in the construction and maintenance of these condensers. The brass plates must be thoroughly annealed; the air must be kept quite dry.

33. POWER CONDENSERS

The condensers used in radio transmitting circuits carry large amounts of power and are called power condensers. They are operated at high voltages, and this requires special construction. The voltage required in transmitter use may be found as follows:

The power input is measured by the energy stored in the capacity at each charge multiplied by the number of charges or discharges per second. Thus $P = \frac{1}{2}CE_0^2N$ where N is the number of discharges per second, or the spark frequency.

$$\text{Hence } E_0^2 = \frac{2P}{CN}$$

In a typical case $N = 1000$, $C = 0.016 \times 10^{-6}$ farad and $P = 2000$ watts; then

$$E_0 = \sqrt{\frac{2(2000)}{1000(0.016)10^{-6}}} = 16000 \text{ volts.}$$

Air and Oil Condensers.—The foregoing indicates the order of the voltage requirement for power condensers. In the case of air condensers of ordinary construction this high voltage would necessitate very large spacing between the plates, and this, in turn, would necessitate a very big volume for a moderate capacity. If, however, the air condensers have a strong air-tight case, so that the air inside the case can be compressed up to 15 or 20 atmospheres, the breakdown voltage becomes very high, even for a small distance between the plates. Thus voltages as high as 35 000 may be used with 3-mm spacing between the plates. At the same time the brush discharge losses are reduced so as to be negligible. Such condensers have the advantage of low power loss, but the disadvantage of being quite bulky. A size commonly used has a capacity of about 0.005 microfarad. Another way of utilizing the air condenser type for high voltages is to fill the condenser with oil. Using ordinary petroleum oils, this about doubles the capacity. Condensers with oil as a dielectric are very well suited for power condensers. The breakdown voltage is very high, dielectric and brush losses low, and on account of the high dielectric constant of some oils, it is easy to get a large capacity in a moderate volume. A variable condenser suitable for oil immersion is shown in Fig. 85, facing page 118. Only condensers with liquid or gas dielectrics can have continuously variable capacity. This is a great advantage, and such a condenser as this is a most valuable supplement to the fixed condensers of solid dielectric used in a high-voltage circuit, since it makes possible easy variation of the wave length.

Glass Condensers.—Condensers with glass for a dielectric, and especially the cylindrical Leyden jars with copper coatings, are very extensively used for power condensers. Special types have been devised to increase breakdown voltage and reduce brush discharge. This is accomplished with the ordinary types by immersion in oil, which also improves the insulation. Glass condensers are relatively cheap. They must, of course, be carefully handled. A stock size of Leyden jar has a capacity of 0.002 microfarad. The power factor of glass condensers is rather large (see next section). When used in the primary circuit of a quenched gap transmitter, however, moderate power loss in the condensers is probably not important in reducing the efficiency of the set since the power losses in the rest of the circuit are high and the circuit is operative, if proper quenching is secured, for but a small fraction of the time. (See Fig. 86, facing p. 119.)

Mica Condensers.—These are coming into use to a considerable extent in radio work, both as power condensers and as standards. They have the advantages of low power loss, small volume, and are not fragile. Stock sizes include 0.002 and 0.004 microfarad. (See Fig. 87, facing p. 119.) In order to withstand high voltages, the mica sheets are comparatively thick and several sections are joined in series. When they are properly made and have very small phase difference, they may be used as standard condensers; and very conveniently supplement the air condensers ordinarily used as standards, being obtainable in larger capacities. They are valuable as standards both on account of their permanence and the large capacity obtainable in a small volume, but must be standardized both in respect to capacity and power factor for the range of frequencies at which they are to be used. While the capacity and power factor may be considerably different at high frequencies from the values at low frequency, it is fortunate that throughout the range of radio frequencies both of these quantities are practically constant. An exception must be made in the case of the power factor if any considerable portion of the power loss is due to ohmic resistance in the leads or plates; for under these circumstances, the power factor will increase with increasing frequency, as pointed out below. In properly constructed condensers, however, power loss from this source is negligible.

34. POWER FACTOR

The power loss in a condenser may be due either to imperfection of the dielectric or to resistance in the metal plates or leads. The dielectric may cause a power loss either by current leakage, by brush discharge, or more commonly by the phenomenon described later under the head of " Dielectric absorption."

Leakage.—The leakage of electricity by ordinary conduction through the dielectric or along its surface contributes to the phase difference at low frequencies but is generally negligible at high frequencies. The effect of leakage on the power factor may be seen as follows: A condenser having leakage may be represented by a pure capacity with a resistance in parallel. The current divides between the two branches, the current I_R through the resistance being in phase with the applied E, and the current I_c through the capacity leading E by 90°. The resultant I leads E by an angle which is less than 90° by the phase difference ψ. From Fig. 88,

$$\tan \psi = \frac{1}{R\omega C}.$$

The effect of R may be shown by an example. A condenser of 0.01 microfarad capacity with an insulation resistance as low

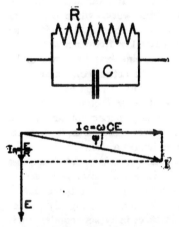

Fig. 88.—*Equivalent circuit and vector diagram for condenser having leakage*

as 10 megohms has, at a frequency of 60 cycles per second, a

$$\text{power factor} = \frac{1}{(10)^7 \, 377 \, (10)^{-8}} = 0.027 = 2.7 \text{ per cent.} \quad \text{This is a}$$

very appreciable quantity; 2.7 per cent of the current flows by conduction instead of by dielectric displacement. This effect, however, decreases as the frequency increases, for the dielectric current increases in proportion to the frequency while the leakage current does not. For instance, at 10 000 cycles, the power factor = 0.00016. Thus at radio frequencies the power factor due to conduction through any but an extremely poor condenser is wholly negligible.

Series Resistance.—A resistance within a condenser in series with the capacity affects the power factor very differently from a resistance in parallel. The series resistance includes the resistance of plates, joints, or contacts, and the leads from binding posts to plates. The E_r across the resistance is in phase with the current I, and the emf E_0 across the capacity is 90° behind I in phase. The power factor $= \sin \psi$, and since ψ is usually small, it may be taken as $= \tan \psi$, which from Fig. 89 is $r\omega C$.

If $r = 1$ ohm and $C = 0.01$ microfarad the power factor at 60 cycles $= 3.8$ $(10)^{-7}$. This is utterly negligible. However, at a frequency of 1 000 000 cycles, the power factor $= 0.063 = 6.3$ per cent, which is so large as to be serious. Thus, it is important to minimize series resistance in condensers for radio work, while condenser leakage on the other hand has its chief importance at low frequencies.

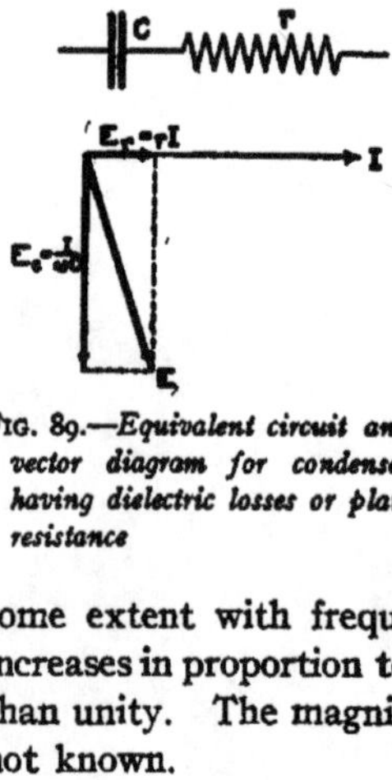

FIG. 89.—*Equivalent circuit and vector diagram for condenser having dielectric losses or plate resistance*

In the foregoing example, r was taken as 1 ohm. In actual condensers it is sometimes greater than this. A plate resistance of several ohms is common in the ordinary paper condenser. In most other condensers, a high series resistance indicates a defect.

On account of skin effect, the series resistance in a condenser increases to some extent with frequency. In consequence, the power factor increases in proportion to a power of the frequency slightly greater than unity. The magnitude of this effect in typical condensers is not known.

Dielectric Absorption.—When a condenser is connected to a source of emf such as a battery, the instantaneous charge is followed by the flow of a small and steadily decreasing current into the condenser. The additional charge seems to be absorbed by the dielectric. Similarly, the instantaneous discharge of a condenser is followed by a continuously decreasing current. It follows that the maximum charge in a condenser cyclically charged and discharged varies with the frequency of charge. The phenomenon is similar to viscosity in a liquid, and is sometimes called "dielectric viscosity."

Dielectric absorption is always accompanied by a power loss, which appears as a production of heat in the condenser. The existence of a power loss signifies that there is a component of emf in phase with the current. The effect of absorption is thus equivalent to that of a resistance either in series or in parallel with the condenser. It is found most convenient to represent absorption in terms of a series resistance, which is spoken of as the "equivalent resistance" of the condenser. An absorbing condenser is, therefore, considered from the standpoint of Fig. 89, and the power factor $= r\omega C$. The equivalent resistance r

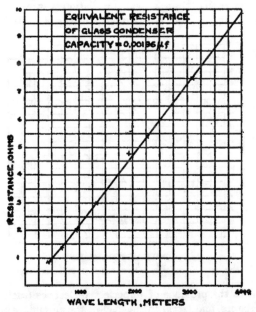

Fig. 90.—*Variation of dielectric loss in glass with wave length*

is constant for a given frequency but is different for different frequencies.

The variation of the power factor and the equivalent resistance with frequency is a complicated matter, the laws of which are not accurately known. To a first approximation, however, the power factor of an absorbing condenser is constant. Since $r\omega C$ is approximately constant, r is inversely proportional to frequency (and therefore directly proportional to wave-length). This is well shown by the nearly straight line in Fig. 90; the crosses show observed resistances for a glass condenser. For most condensers used in radio circuits, the power factor is constant over the range of radio frequencies and has nearly the same value at low frequencies. This fact, viz, that power factor or phase difference is approximately independent of frequency, is very convenient and easily remembered.

The same law holds for antennas at frequencies less than those for which the radiation resistance is large; the equivalent resistance is inversely proportional to frequency. Otherwise expressed, the equivalent resistance is proportional to wave length, for wave lengths greater than those at which radiation is appreciable. This is shown by Fig. 91, and is discussed on page 81. It is believed

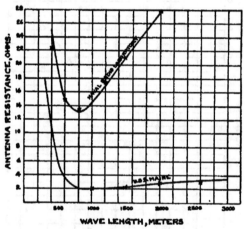

FIG. 91.—*Variation of antenna resistance with wave length*

to be due to the presence of imperfect dielectrics (such as buildings, insulators, and trees) in the field of the antenna; thus an antenna is like an air condenser in respect to the effect of poor dielectrics in its field, as well as in other ways.

Values of power factor and equivalent resistance are given for typical radio condensers in Table 2. The power factor does not vary with the size of the condenser; it is a function of the dielectric and not of the particular condenser. The equivalent resistance, on the other hand, is inversely proportional to the capacity of the condenser, since power factor $= r\omega C$. The data in the table for Leyden jars in oil are really typical of the glass dielectric, while the relatively large power factor given for Leyden jars in air is due to brush discharge. The brush discharge is a function of voltage (see Brush Loss below), and is only appreciable at voltages above 10 000. For all voltages below this, the power factor of a Leyden jar in air is about the same as the value given for a Leyden jar in oil, about 0.003.

TABLE 2.—Power Factors of Radio Condensers [a] (at 14 500 volts)

Kind of condenser	Power factor	Capacity	Equivalent resistance at 1000 m
Compressed air [b]	0.001	0.0058	0.14
Glass (Leyden jar) in oil	.003	.0060	.28
Molded (Murdock), new	.004	.0054	.41
Glass plates in oil	.005	.0042	.58
Glass (Moscicki type)	.006	.0055	.57
Glass (Leyden jar) in air	.016	.0061	1.4
Molded micanite	.023	.0041	2.9
Paper	.024	.0058	2.2

[a] Calculated from determinations by Austin, Bull. B. S., 9, p. 77; 1912.

[b] The power factor observed for the compressed-air condenser is attributable not to the air but to the resistance in its leads and plates, to eddy current loss in the metal case, and to the insulating material used to separate the two sets of plates.

Except for the compressed-air condenser and the Leyden jar in air, the power factors do not vary with voltage nor to a great extent with frequency. The power factors of solid dielectrics are principally due to dielectric absorption. If an appreciable portion of the power factor were due to a series resistance, such as a high plate resistance, it would be manifested by an increase of power factor with increasing frequency. At frequencies higher than 300 000 (corresponding to a wave length of 1000 meters), this is to be expected in the case of paper condensers, which have long tin-foil plates; the power factor will increase, and the equivalent resistance can not decrease to a value lower than the plate resistance. When the power factor is due purely to dielectric absorption the equivalent resistance decreases in proportion to an increase of frequency (or increases in proportion to an increase of wave length).

The dependence of power loss, power factor, and equivalent resistance upon the frequency, in the case of absorbing condensers, condensers with leakage, etc., is summarized in the following table. P is the power loss, r the equivalent series resistance, ψ the phase difference (same as power factor), ω the usual 2π times frequency, and λ wave length. Much of the information in the table is contained in the equations

$$P = \omega C E^2 \psi$$

$$\psi = r\omega C$$

The table shows the powers of frequency and wave length to which P, ψ, and r are proportional. An arbitrary notation is used in the exponents of ω and λ; the symbols $>$ and $<$ for "greater than" and "less than" are used to indicate the magnitudes of the powers.

TABLE 3.—Variation of Phase Difference, etc., with Frequency

Kind of system	P		ψ		r	
Condenser with leakage.......	ω^0	λ^0	ω^{-1}	λ^1	ω^{-2}	λ^2
Absorbing condenser at very low frequencies.............	$\omega^{<1}$	$\lambda^{>-1}$	$\omega^{<0}$	$\lambda^{>0}$	$\omega^{<-1}$	$\lambda^{>1}$
Absorbing condenser at radio frequencies.................	ω^1	λ^{-1}	ω^0	λ^0	ω^{-1}	λ^1
Constant series resistance.....	ω^2	λ^{-2}	ω^1	λ^{-1}	ω^0	λ^0
Skin effect in conductors......	$\omega^{>2}$	$\lambda^{<-2}$	$\omega^{>1}$	$\lambda^{<-1}$	$\omega^{>0}$	$\lambda^{<0}$

Brush Loss.—When a condenser is operated in air at high voltage, more or less ionization occurs at the edges of the plates. When the edges of the plates are entirely exposed to the air, as in a Leyden jar, this effect is large and a considerable power loss occurs at high voltages. At very high voltage the discharge is so great as to be evident in the form of a visible brush from the edges of the conducting plates, and is especially strong at corners and points. At such voltages the power factor due to the brush discharge is large compared with that caused by absorption and other causes.

Austin [15] has found that the power factor and, hence, the equivalent resistance of a Leyden jar in air do not vary with voltage up to about 10 000 volts; between 10 000 and 22 000 volts they increase approximately as the square of the voltage; and at voltages greater than 22 000 increase faster than the square of the voltage. Since the power loss is proportional to the square of voltage and to the power factor, a constant power factor means power loss proportional to the square of the voltage. Thus, the

[15] See reference No. 111, Appendix 2.

power loss in Leyden jars below 10 000 volts is proportional to the square of the voltage. Between 10 000 and 22 000 volts, where brushing occurs and the power factor is proportional to the square of voltage, it follows that the power loss varies as the fourth power of the voltage. Above 22 000 volts the power loss increases faster than the fourth power of voltage.

35. MEASUREMENT OF CAPACITY

A capacity measurement at radio frequencies is usually a comparison of the unknown capacity with a standard variable condenser. The most precise measurements are made by means of a direct substitution. A primary standard for such work is a condenser which has been calibrated at low frequency and is so designed as to have extremely low absorption and high insulation resistance, so that its capacity is the same at radio and at low frequencies. For ordinary measurements it is possible to use a condenser with moderate absorption and insulation resistance which has been compared at radio frequencies with a primary standard. A condenser with high absorption or low insulation resistance may show changes in capacity in the range of radio frequencies, will increase the resistance of the circuit into which it is introduced, and is not suitable for a standard. The inductance of the leads within a standard condenser should be small in comparison with that of the coil, or the apparent capacity will vary with the frequency.

The unknown condenser may be fixed or variable. It is inserted in series with an inductance coil in a circuit which is provided with a device to indicate resonance; that is, a wavemeter circuit. The source of oscillations is loosely coupled to this circuit and the frequency of the source adjusted until resonance is obtained. The unknown condenser is then removed and the variable standard substituted in its place and the setting of the standard found for which resonance is again obtained. The capacity of the standard at this setting is equal to that of the unknown.

In order to attain a high accuracy in the comparison, certain precautions must be observed. The capacity of the rest of the circuit to each of the two condensers must be either very small or the same. On this account it is desirable that the leads to the condenser be fairly long and fixed. Proximity of the metal shields, etc., of the condensers to the coil will reduce its inductance on account of eddy currents, and since this effect may vary with the two condensers, it is desirable to eliminate it by the use of long leads. In order that the capacities should be definite, both condensers should be shielded, the shield connected to one condenser terminal and connected to ground, the same lead from the inductance coil being connected to the earthed terminal in both cases.

When the unknown condenser is also a variable and is to be calibrated at a number of points, it is convenient to use a throw-over switch to change from one condenser to the other. In this case it is necessary that the inductance and capacity of the two pairs of leads running from the switch to the condensers and mutual capacities to the rest of the circuit should be very nearly the same in the two positions. Any error from this cause may be checked by using the same condenser first on one set of leads and then on the other, and comparing the settings for resonance.

Calibration of Large Variable Condenser.—A large variable may be calibrated against a much smaller standard variable in the following manner. The large unknown is first set at a low setting and directly compared with the standard as outlined above. This gives the capacity of the large condenser at the low setting with a high accuracy. Then the two condensers are connected in parallel, the terminals of the two condensers which are connected to the shields being connected together. The large condenser is set at the known point and the small condenser at its maximum capacity and the source is adjusted to resonance. Then keeping the frequency of the source unchanged, the setting of the large condenser is increased by the desired steps, while the setting of the standard is reduced to compensate. From the reduction in capacity of the standard, the increase in capacity of the large condenser from that at the known point can be computed. When the standard can be no longer reduced, the large condenser is set at the highest determined value, the standard is again set at its maximum, and the wave length of the source increased until resonance is again obtained. The process of increasing the capacity of the unknown and decreasing that of the standard is then repeated.

Effect of Internal Lead Inductance.—If the unknown condenser has internal leads which have appreciable inductance, its apparent capacity will increase as the wave length is decreased. To determine this inductance the condenser is measured at a very long wave length with a large coil where the effect of the leads will be negligible and then with a small coil of inductance L. Calling the capaçity at long wave lengths C_1 and the apparent capacity at short wave lengths C_2 and the inductance of the leads l we may write

$$C_2L = C_1(l+L)$$

or

$$l = \frac{C_2-C_1}{C_1}L$$

This method requires the use of a standard condenser with leads of negligible inductance.

COILS

36. CHARACTERISTICS OF RADIO COILS

To introduce a certain amount of inductance into a circuit, a coil of copper strip or specially stranded wire is ordinarily used. This

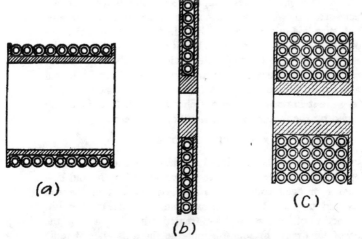

(a)

(b)

(C)

FIG. 92.—*Typical forms of inductance coils*

is usually mounted or wound on a form made of such insulating materials as wood or bakelite or other composition. The form is usually hollow, and in some cases the conductor is supported only by strips of insulator at intervals. The turns of wire or strip are usually circular, but sometimes are polygonal. Iron is not used as a core in inductance coils for radio use, because such coils have high effective resistance, the power losses increasing with frequency. Three types of coil winding are widely used in radio work; the single layer, the flat spiral or pancake, and the multiple layer. Cross sections of these windings are shown in (a), (b), and (c) of Fig. 92.

The types (a) and (b) are universally used for coils of low or moderate inductance, the type (b) is especially convenient for portable instruments on account of its compactness. The multiple-layer coil is generally used where it is desired to obtain a large inductance in a compact form.

The important electrical characteristics of a coil are its inductance, resistance, and capacity. In Part III are given formulas, with tables and examples, which cover the calculation of the inductance of practically all types of coils which are used in radio circuits.

On account of the change of current distribution within a conductor, the self-inductance tends to decrease as the frequency increases. There is no appreciable change of mutual inductance with frequency. The change of self-inductance is very small, and can be calculated only in a few simple cases. The inductance of a coil decreases somewhat more than the inductance of the same length of wire laid out straight. The effect of distributed capacity, however, is to increase the inductance. At low frequencies the design of coil form is usually determined by the requirement of minimum resistance with a given inductance. This is treated on page 287. At radio frequencies, however, the capacity of coils is of very great importance and the choice of coil form is largely determined by the requirement of small coil capacity.

37. CAPACITY OF COILS

In section 19 it has been pointed out how the capacity of coils may lead to circuits which resonate to two wave lengths and how overhanging or dead ends of coils may, on account of the coil capacity, seriously affect the reactance and resistance of a circuit. In this section other cases will be treated which show how the capacity of coils may considerably increase the resistance of radio circuits.

Effect of Capacity on Inductance and Resistance.—In the first case it is assumed that an emf (see Fig. 93(a)) is introduced into a circuit by means of a coupling coil L' of few turns of negligible inductance and not by coupling with the main coil L. Under these conditions the capacity of the coil L will affect both its resistance and apparent inductance and is not merely added to that of the condenser as would be the case if the coupling coil were removed and the emf introduced into the circuit by induction in the coil L itself. Instead, the coil L and its capacity form a parallel circuit as shown in Fig. 93(b).

If we write $Z = \sqrt{R_a^2 + \omega^2 L_a^2}$, where R_a and L_a are the apparent resistance and inductance of the parallel circuit, it may be shown that

$$R_a = \frac{R}{\omega^2 C_0^2 R^2 + (1 - \omega^2 L C_0)^2} \tag{70}$$

$$L_a = \frac{L(1 - \omega^2 L C_0) - C_0 R^2}{\omega^2 C_0^2 R^2 + (1 - \omega^2 L C_0)^2} \tag{71}$$

The terms $\omega^2 C_0^2 R^2$ and $C_0 R^2$ are very small and negligible, excepting when very close to the frequency for which $(1 - \omega^2 L C_0)$ equals zero. This is the frequency with which the coil L would oscillate by itself—that is, closed only by its own capacity. At this fre-

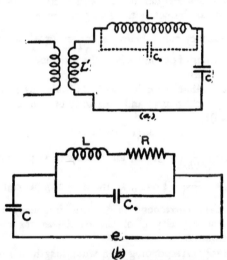

Fig. 93.—*Equivalent circuit of a coil having distributed capacity placed in series with a condenser*

quency. these terms determine the resistance and inductance of the coil. For other frequencies we may write

$$R_a = \frac{R}{(1 - \omega^2 L C_0)^2} \tag{72}$$

$$L_a = \frac{L}{(1 - \omega^2 L C_0)} \tag{73}$$

From these equations we see that the resistance and inductance start at the values R and L at low frequencies ($\omega = 0$) and increase as the frequency increases, the resistance increasing about twice as rapidly in per cent as the inductance. When approaching the frequency $\omega = \dfrac{1}{\sqrt{LC_0}}$ the resistance becomes very great and beyond this frequency falls off again, finally becoming zero. The inductance also becomes very great, but just before the frequency $\omega = \dfrac{1}{\sqrt{LC_0}}$ is attained we see from equation (71) that it falls off very rapidly and becomes highly negative. For higher frequencies it remains negative, but approaches zero as the frequency is still further increased. This negative value of inductance means that the coil is behaving as a condenser, and that for the circuit to resonate to such frequencies the condenser C would have to be replaced by an inductance. The primary interest, however, lies in the frequency range below $\omega = \dfrac{1}{\sqrt{LC_0}}$ for which range it will be shown by means of examples how the resistance and inductance of L are increased.

Let us assume that the coil L with its leads has an inductance of a millihenry (10^{-3} henry) and a capacity of 50 micromicrofarads (5×10^{-11} farad).

If $\omega = 10^6$, then $(1 - \omega^2 L C_0) = (1 - 10^{12} \times 10^{-3} \times 5 \times 10^{-11}) = 0.95$

and
$$R_a = \frac{R}{(0.95)^2} = 1.11\, R, \text{ an increase of 11 per cent.}$$

$$L_a = \frac{L}{(0.95)} = 1.05\, L, \text{ an increase of 5 per cent.}$$

The wave length corresponding to this frequency is $\lambda = 1885$ meters, and the capacity C of the condenser for resonance is 950 micromicrofarads.

If $\omega = 1.5 \times 10^6$ corresponding to a wave length of 1250 meters, the resistance increase is 26 per cent and the inductance increase 12 per cent. The capacity C would then be 400 micromicrofarads.

These are examples of conditions that may readily occur in practice. It is very desirable in designing radio circuits to arrange whenever possible that the emf be introduced in the main coil L itself.

Effect of Dielectric Absorption in Coil Capacity.—Even in the case of a simple oscillatory circuit of coil and condenser with the emf applied by induction in the coil (in which case the coil capacity may be considered as in parallel with that of the condenser) the resistance of the circuit may be increased. This is not due merely to the capacity of the coil, but because the coil capacity is, in general, a highly absorbing condenser. The dielectric is the insulation of the wire and the material of the coil form, and the resulting phase difference of the coil capacity may be several degrees. The resistance added to the circuit on this account will be larger the greater the absorption of the coil capacity, and the larger this capacity relative to the condenser capacity, and also the longer the wave length.

To illustrate the importance of this effect, let us take the same data as used in the preceding examples, with the further assumption that the coil capacity C_0 has a phase difference of $2°$.

In the first example $\omega = 10^6$, $C_0 = 50$ micromicrofarads, and $C = 950$ micromicrofarads. The phase difference of C_0 and C in parallel will be $\dfrac{50}{950 + 50}$ times $2° = 6'$. The effective resistance added to the circuit on account of absorption will be

$$\rho = \frac{\tan \theta}{\omega(C + C_0)} = \frac{\tan 6'}{10^6(10^{-9})} = 1.75 \text{ ohms.}$$

In the second example, $\omega = 1.5 \times 10^6$, $C_0 = 50$ micromicrofarads, $C = 400$ micromicrofarads. Hence, $\theta = 13.'3$

$$\rho = \frac{\tan 13.'3}{1.5 \times 10^6 (0.45 \times 10^{-9})} = \frac{3.88 \times 10^{-3}}{0.67 \times 10^{-3}} = 5.8 \text{ ohms.}$$

It is therefore clear that it is extremely important so to design coils that the capacity shall be small and of as low absorption as possible.

Reduction of Coil Capacity.—The coils of type (a) and (b) have low capacities, while type (c) has a large capacity. This results from the fact that in a coil of type (a) or (b) with n turns, only $\dfrac{1}{n}$ th of the total voltage acting upon the coil is impressed between the adjacent turns, while in the multiple-layer coil as ordinarily wound, turns which have high voltages between them will be adjacent. For example, in a coil of two layers, if one layer is first wound, and then the other on top of it, the first and last turns of the coil will be adjacent and the total voltage on the coil will be impressed between them. The capacity of such a coil introduces serious errors in high-frequency measurements unless the coil is accurately standardized.

The capacity of coils of type (a) or (b) can be reduced and the phase difference improved if the windings are separated slightly with air between. On account of the high voltage between the adjacent turns this is a customary procedure in the case of coils designed to carry large currents, such as the familiar flat spiral coils used in transmitters. In the standard wave-meter coils of this Bureau this method of reducing the capacity and phase difference has been used. The coils are shown in Fig. 215, facing page 318, and range in inductance from 60 to about 5000 microhenries. The coil capacities, including fairly long leads, range from 9 to 16 micromicrofarads. No data are available as to the phase differences due to these coil capacities, but since the windings are separated from each other with air between, excepting at the points of support on the coil form, the phase differences are extremely low.

In coils of type (c), Fig. 92, the capacity may be reduced by using the so-called "banked" winding. Instead of winding one layer complete and then winding the next layer back over the first,

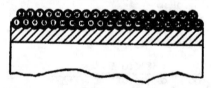

FIG. 94.—*Method of banking the winding of a coil in order to reduce its distributed capacity*

one turn is wound successively in each of the layers, the winding proceeding from one end of the coil to the other. The best results are obtained in a coil of a few layers. The method is illustrated in Fig. 94 for a coil of two layers, the succession of the turns being indicated by the numbers. The turns in the lower layer are prevented from slipping during the winding by grooving the coil form or covering it with rubber tape. The maximum voltage between adjacent wires in this two-layer coil is the voltage corresponding to three turns.

38. MEASUREMENT OF INDUCTANCE AND CAPACITY OF COILS

When the inductance of a coil is measured at high frequencies, the quantity obtained is the apparent inductance; that is, a quantity in which is combined the effects of the pure inductance and the capacity of the coil. (See p. 132.) Thus it is customary to speak of the inductance of a coil at a given wave length or frequency. The inductance of a coil at a given frequency is measured by comparing it with a standard coil or by determining the capacity required in parallel with the coil to form a circuit which is resonant to the known frequency.

In the first method let L_s be the inductance of the standard coil and L_x that of the unknown. The source is set at a given frequency and a circuit containing L_s and a condenser is tuned to resonance with the source and the condenser setting read. Let the capacity be C_s. Then L_x is substituted and resonance again obtained, the capacity being C_x. Since the frequency is the same in both cases we have

$$L_s C_s = L_x C_x$$

$$L_x = \frac{L_s C_s}{C_x}$$

Here L_s is the apparent inductance of the standard coil at the frequency of the measurement. If the pure inductance L_p and capacity C_o of the standard coil is given

$$L_p (C_s + C_o) = L_x C_x$$

$$L_x = \frac{L_p (C_s + C_o)}{C_x}$$

In the second method, the frequency or wave length furnished by the source must be accurately known. The circuit with the unknown coil L_x is adjusted to resonance with the source. If the resonant capacity is C_x we have

$$\lambda = 1.885 \sqrt{L_x C_x}$$

$$L_x = \frac{3.553 \; \lambda^2}{C_x}$$

where λ is wave length in meters.

C_x is capacity in micromicrofarads.

L_x is inductance in microhenries.

Determination of Coil Capacity.—The pure inductance and capacity of the coil may be derived from the observed values of apparent inductance in various ways. A simple method makes use of the formula given on page 133 for the apparent inductance in terms of the pure inductance and coil capacity. Denoting two frequencies or capacity settings by subscripts 1 and 2 we have

$$L_{s1} = L\left(1 + \frac{C_o}{C_1}\right)$$

$$L_{s2} = L\left(1 + \frac{C_o}{C_2}\right)$$

or

$$\frac{1 + \dfrac{C_0}{C_1}}{1 + \dfrac{C_0}{C_2}} = \frac{L_{a_1}}{L_{a_2}}$$

or approximately

$$1 + \frac{C_0}{C_1} - \frac{C_0}{C_2} = \frac{L_{a_1}}{L_{a_2}}$$

and

$$C_0 = \left(\frac{L_{a_1}}{L_{a_2}} - 1\right)\frac{C_1 C_2}{C_2 - C_1}$$

Knowing C_0, L may be obtained by substitution.

If one determination is made with a very large condenser capacity and another with a small condenser, we may write

$$L_{a_1} = L \text{ approx.}$$

$$L_{a_2} = L\left(1 + \frac{C_0}{C_2}\right)$$

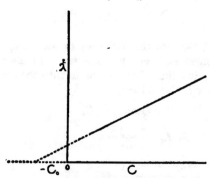

Fig. 95.—*Graphical method of determining the capacity of a coil*

and

$$C_0 = C_2\left(\frac{L_{a_2}}{L_{a_1}} - 1\right)$$

Another method which is likewise of considerable value in wave-meter calibrations makes use of a plot of the wave length squared against the condenser capacity. Since

$$\lambda = 1.885 \sqrt{L\,(C + C_0)}$$

$$\lambda^2 = 3.553\, L\,(C + C_0)$$

Since L, the pure inductance, is a constant, the relation between λ^2 and C is linear and the plot will be a straight line as in Fig. 95. The slope of this line $\dfrac{d\,(\lambda^2}{d\,C} = 3.553\,L$ determines the pure inductance of the coil. The distance from the origin to the intercepts of the line with the axis of abscissæ determines the capacity of the coil, for when $\lambda^2 = 0$; $C = -C_0$. Another method of determining the coil capacity making use of the harmonic oscillations of a pliotron generator has already been described in the section on wavemeters, page 100.

CURRENT MEASUREMENT

39. PRINCIPLES

Current measurements at radio frequencies are made in transmitting sets, in wave meter and other testing circuits, and in receiving sets. The instruments used in such measurements, their principles, and the precautions necessary in using them, are discussed in this section. The measurement of current is a cardinal operation in high-frequency work, to a much greater degree than at low frequencies, since upon it depends also the measurement of resistance, and it is involved in most of the measurements of other quantities. There are three distinct ranges of currents: Large currents such as are used in large transmitting sets; moderate currents in small sending outfits and in wave meters, decremeters, etc., used at a sending station; and the very small currents in receiving circuits and in testing equipments operated by buzzers and other weak sources. These three ranges will be treated separately, beginning with the middle range of moderate currents, because the simplest instruments are there used and the fundamental principles can best be discussed in connection with them.

Most of the ammeters used for low-frequency currents are entirely unsuitable at high frequency. Most of them are ruled out by the requirement[16] that the circuit within a high-frequency ammeter must be of the greatest simplicity, because its inductance and its capacity must be as small as possible. Large impedance and large capacity would tend to make the current flow through the dielectric as well as through the conductor, in amount varying with the frequency. This would cause the readings of the instrument to change with frequency. The require-

---

[16] See reference No. 195, Appendix 2.

ment of a simple form of circuit is best fulfilled by a single straight wire of very small diameter. The current flowing in such a conductor is most simply measured by its heating effect, and hence it has come about that most ammeters for high frequency have been of the hot-wire or thermal type. Other types have been used to a limited extent.

The Thermal Ammeter.—The basic principle of the thermal ammeter is given by

$$P = RI^2,$$

P being the power consumed as heat in the instrument. The deflection of the instrument depends upon the heating effect P, which is indicated in one of the various ways discussed below. In order that a given deflection should always correspond to the same current it is necessary that the relation of P to I should remain constant, and this requires that R, the resistance of the instrument, should not change with frequency. In order not to vary in resistance at radio frequencies, a conductor must have very small cross section (see section below on Resistance). Hence, the working element of a thermal ammeter must be a fine wire or a very thin strip of metal. For the accuracy required in most radio work it is only necessary that the resistance of the wire or strip should not change by more than 1 per cent over any range of frequency used. The largest wire which may be used in an ammeter is thus found by consulting Table 18, page 310, for the highest frequency to be used. In ammeters which are intended for use at the highest frequencies of radio practice, viz, about 2 000 000 (corresponding to a wave length of 150 meters), the largest diameter permissible for a copper wire is 0.08 mm and for a constantan wire is 0.4 mm. Using a single wire of such a size, only a few amperes can be measured, because larger currents would overheat the wire and alter its properties.

The length of the wire is not important; it must simply be long enough so that the current distribution within it is not appreciably altered by the terminals to which it is connected. A length of 10 cm or more is satisfactory. If properly constructed, such instruments may be calibrated with direct current or low-frequency alternating current and the calibration assumed correct at high frequency. It is always safest, however, to calibrate the ammeter by comparison with some recognized standard at the frequencies at which it will be used.

Sources of Error.—Certain precautions must be observed in the use of high-frequency ammeters; some of these will be pointed out below in connection with particular types of instruments. There is one source of error to which all are subject, whether thermal ammeters or not, at extremely high frequencies. For very short wave lengths, the current is usually not the same at all points of a circuit. The capacity between parts of the circuit and the surroundings is so important that an appreciable fraction of the current is shunted through the dielectric and so different amounts of current flow in different parts of the wire circuit. At such frequencies one must be careful to place the ammeter at that point of the circuit where the value of the current is desired. In order to diminish the flow of current away from the circuit, special devices are used in some ammeters to prevent grounding of the circuit to the ammeter case. For example, where the deflection of the instrument depends on thermal expansion a short length of insulation may be placed in the indicating wire; and in instruments depending on the heating in a thermocouple, the heater and the thermocouple may be separated instead of in contact.

40. AMMETERS FOR SMALL AND MODERATE CURRENTS

The ammeters used for measuring radio currents of about 0.003 to 3 amperes are hot-wire instruments of the simplest form. They consist essentially of a single fine wire, with a means of indicating the heat produced. The heat production may be indicated by any thermometric method, and the following are in use: Expansion, thermoelectric effect, and calorimetry. These three types of instrument are discussed in the following.

Current-Square Meter.—The most familiar instrument of the expansion type is the current-square meter, used in wave meters and decremeters. This has sometimes been called a "wattmeter," the deflections being proportional to the watts consumed within the instrument itself; such a name is utterly misleading and undesirable. The hot wire is of resistance metal, and its increase in length when heated is indicated by a pointer operated by a thread attached to the wire and held taut by a spring. The scale usually has graduations equally spaced, so that the readings are proportional to the square of the current. This makes the instrument particularly useful in measurements of resistance by the reactance variation method. The scale can, of course, be graduated so as to read current directly, but such a scale is badly crowded at the lower end.

A familiar type of this instrument gives full-scale deflection with 0.08 ampere. The instrument is shown in Fig. 96, facing page 156. It has a resistance of about 5 ohms. If intended for use in actual measurements of current the instrument can not be shunted, for reasons explained in the next section. For use in a wave meter or decremeter, however, it may be shunted to give whatever current range may be desired, since in such use the variations of frequency are so small during any one measurement that the shunt does not affect the accuracy of the result.

A source of serious error in the use of these instruments is the presence of electrostatic charges on the glass covering the dial. If the glass happens to be stroked with any object the pointer may deflect several scale divisions and remain in the new position a long time. This is not an error which can be eliminated

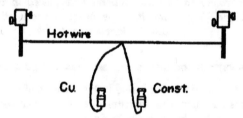

FIG. 97.—*Simple hot-wire ammeter with thermoelectric indicating device*

by using the zero adjustment, because the electrostatic charges on the glass exert a force on the pointer which varies with the position of the pointer. The most convenient way to eliminate this effect is to breathe on the glass, the charge being conducted away by the layer of moisture. This effect is especially troublesome in cold weather, when electrostatic charges are readily produced and maintained.

All hot-wire instruments show more or less zero shift. After current has been flowing through the instrument, the pointer does not return exactly to its original position. In a current-square meter having a zero adjustment this is usually overcome by first allowing current to flow for several seconds without taking a reading, then cutting off current and setting zero, and then allowing current to flow and reading. If the instrument has no zero adjustment, error may be eliminated by first allowing current to flow without taking a reading, then cutting it off and reading zero, then reading current, and then reading zero again. The mean of the two zero readings is to be subtracted from the current reading. (In instruments having a nonuniform scale the zero reading should be made as explained on p. 149.)

Thermoelectric Ammeter.—The heat developed in the hot wire may be indicated by means of a thermocouple, placed very near or in contact with the wire. The electromotive force produced by the heating of the thermocouple is measured by a suitable direct-current instrument. The indications depend upon the temperature at one point only of the hot wire instead of upon the heating effect throughout the whole wire as in the expansion ammeter. There are two types in use for moderate currents. A diagram of the simple type is shown in Fig. 97. *A* is the fine wire which carries the high-frequency current. The copper-constantan thermocouple is hard soldered to *A* and connected to the binding posts and thence to a galvanometer. Such an instrument is easily constructed for laboratory use for currents up to 2 amperes. Commercial instruments are made by combining the hot wire, thermocouple, and a pointer-type microammeter into a single instrument.

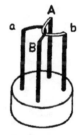

Fig. 98.—*Crossed-wire type of thermoelectric ammeter*

The other type of thermoelectric ammeter is the crossed-wire type, illustrated in Fig. 98. Two fine wires, *Ab* and *aB*, of constantan and copper, constantan and platinum, or similar pair of metals, are in contact at one point, crossing each other as shown. Connection is made from the points *A* and *B* to the source of high-frequency current. The current heats the fine wires, raising the temperature of the junction, and this causes an emf between *a* and *b*, to which points a microammeter or galvanometer is connected. The deflections of the usual thermoelectric ammeters are approximately proportional to the square of the measured current.

Thermoelectric ammeters should be calibrated with low-freqency alternating rather than direct current, for the reasons stated on page 170, below.

Air-Thermometer Ammeter.—This instrument, which is really a calorimeter, was formerly used in wave meters and in measurements of high-frequency resistance, for currents of a few hundredths to a few tenths ampere. Current passing through the fine wire (Fig. 99) heats the air in the glass bulb and causes the alcohol in the right side of the U-tube to rise. In order to eliminate the effect of other heat than that produced in the hot wire it is desirable to inclose the bulb in a vacuum jacket. A null instrument is easily made on the air-thermometer principle, by connecting a bulb to the right side of the U-tube, entirely similar to the one shown on the left side. If the two sides are exactly

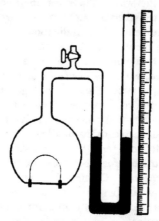

FIG. 99.—*Air-thermometer ammeter*

alike, the liquid column will not move when the RI^2 in one wire equals the RI^2 in the other. These instruments are no longer used in practice.

41. THERMAL AMMETERS FOR LARGE CURRENTS

It is a common dictum that ammeters used in measuring alternating currents of even rather low frequency must not be shunted. This is obvious from consideration of the expression relating the current I_1 in the shunt and I_2 in the instrument, neglecting mutual inductance,

$$\frac{I_1^2}{I_2^2} = \frac{R_2^2 + \omega^2 L_2^2}{R_1^2 + \omega^2 L_1^2},$$

since usually the inductances of the instrument and the shunt are in a different ratio from that of the resistances, and hence the distribution of the current varies with frequency. In fact, even if $\frac{L_2}{L_1} = \frac{R_2}{R_1}$, which would make the expression for $\frac{I_1^2}{I_2^2}$ independent of ω, the current distribution in an actual case is likely to vary with frequency because of the mutual inductance. Now, any high-frequency ammeter in which the circuit within the ammeter itself consists of more than a single elementary filament—or its closest approximation, a fine wire—in reality involves shunting, and needs most careful consideration before it can be pronounced free from error. The proper design of ammeters suitable for measuring radio currents up to several hundred amperes is therefore a difficult matter.

Parallel Wires or Strips.—When larger currents than about 3 amperes have to be measured, the single wire will not suffice; since the wire must be of so small a diameter that its resistance is not changed by frequency, large currents will overheat it. To measure large high-frequency currents, therefore, the current must have more than one path, and it is common to use either wires or very thin strips of metal in parallel. This amounts to shunting, and it is therefore difficult to make these ammeters accurate at radio frequencies. Many of those in use have large errors. The heating effect is indicated either by an expansion device or by a thermocouple attached to one of the wires or strips. If the ratio of the current in that particular wire or strip to the total current changes with frequency, the instrument will be proportionately in error. The errors are much larger than they would be if the indicated current depended on the heat production of the whole current, since the changes of current distribution within a particular system are changes of the first order of magnitude compared to which the change of heat production in the whole system is of the second order.[17]

The simplest arrangement of parallel elements which has been used in this class of ammeters is a group of fine wires or thin strips all in a single plane, as in Fig. 100. The arrangements shown in these diagrams are subject to several errors. In the first place, the self-inductances of the lugs to which the wires or strips are fastened, while utterly negligible at low frequencies, contribute a large part of the impedance at high frequencies. Because of this, the current in the wire z (Fig. 100) will decrease relative to the current in the wire u as the frequency increases. This source of error may be eliminated by connecting the current leads at opposite corners, instead of the two adjacent corners as shown.

---

[17] See reference No. 195, Appendix 2.

Effect of Mutual Inductances.—A more important source of error is the mutual inductance between the wires or strips. The mutual inductance between x and z is less than the mutual inductance between x and y, Fig. 100. The combined effect of all the mutual inductances is that the reactance of the outer wires is less than of the inner wires. The resistances of the wires determine the distribution of current at low frequencies, but as the frequency is increased the reactances become more and more important. Consequently, at high frequencies more current flows in the outer wires than in those near the center. This effect is greater the closer the wires or strips are spaced. If the indicating device (expansion arrangement or thermocouple) is placed on one of the outside wires or strips, the ammeter reads high at high frequencies,

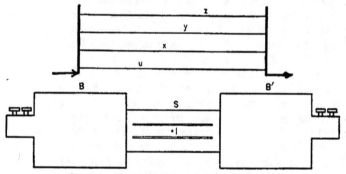

FIG. 100.—*Ammeters with heating elements in parallel*

and if on one near the center of the group, it reads low at high frequencies.

The errors due to mutual and self inductance can be avoided by arranging the wires or strips as equidistant elements of a cylinder and leading the current in to the centers of the ends of the cylinder. The self-inductance is the same in each current path, and each has the same set of mutual inductances with respect to the others. The currents in the different paths must then be the same at high frequencies. It does not follow that such an instrument is free from error, for the reason that the resistances of the individual wires or strips may be quite different because of variations of hardness and small variations of cross section. It is very difficult in practice to get such thin wires or strips of uniform thickness. The inductances do not vary appreciably with variation of cross section while the resistances do of course, differ on this account. Thus the current distribution among the several current paths is not uniform at low frequencies, where resistance is the determining factor, while it is uniform

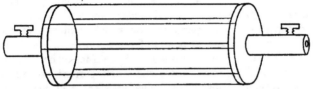

Fig. 101.—*Ammeter with the parallel heating elements arranged cylindrically*

at frequencies so high that only the inductances determine it. Ammeters of this type are used for measuring currents up to 300 amperes. A commercial ammeter of this type is shown in Fig. 102, facing page 156. They are, in general, satisfactory, but the readings change with frequency by a few per cent for the reason just explained. The residual errors in the cylindrical type of ammeter may be practically eliminated by the use of a thermocouple on each wire or strip.

Advantage of High-Resistance Elements.—It is an interesting fact that the changes in current distribution in thermal ammeters occur just in the range of radio frequencies. This is the reason why it is difficult to design such instruments properly. The size of the conductors used (wires having diameters of the order of 0.1 mm and strips of that thickness) is such that their inductive reactance is negligible compared to their resistance below a frequency of about 100 000 but is much greater than the resistance above 2 000 000, so that the current distribution depends on very different properties outside these limits and passes through a variation within this range. The greater the resistance of the conductors the higher are these limiting frequencies. By using high-resistivity wires or strips, therefore, the errors of these instruments can be moved up to rarely used frequencies. There is a limit to this use of high resistance, in that the increase of resistance means that the conductor gets hotter, or can carry less current for a given temperature rise. Platinum-iridium or platinum-rhodium are desirable materials in view of the requirements, having moderately high resistivity and being capable of standing high temperatures. With very fine wires or very thin strips of these materials, well separated but not necessarily arranged in the cylindrical form, ammeters for currents up to 20 amperes are reliable for all except the highest radio frequencies.

So-Called Unshunted Ammeter.—A type of thermal ammeter which has been used extensively to measure radio currents up to 10 amperes is mentioned here only to point out its utter unsuitability. This is the so-called unshunted ammeter, represented diagrammatically in Fig. 103. *A* and *B* are the current leads. They connect to thick copper bars, from which flexible silver strips take the current to several points of the hot wire, whose expansion is measured by the ordinary device (not shown in Fig. 103). Thus a single wire carries the whole current, and the instrument is called unshunted. The resistances of the copper bar and metal strips are negligible in comparison with the resistance of the hot wire, and if the lengths of the sections in parallel are the same, each carries the same current as any other on low

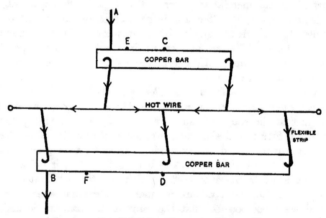

FIG. 103.—*So-called unshunted ammeter of four sections*

frequency. But the inductances of these parts are by no means negligible, and consequently on high frequency the different portions of the wire carry different amounts of current. In fact, in practical cases the impedance of the hot wire itself is but a small part of the impedance of each current path for high frequencies. Even the mutual inductances of the different portions of the "hot wire" are not negligible and in themselves tend to cause more current to flow in the central sections than in the outer sections. The changes in current distribution from the uniformity of direct-current distribution are, in fact, very large. They are equivalent to an increase in the resistance of the system as a whole, so that these instruments read high on high frequency. Whenever an accuracy of 10 per cent or better is desired, this type of ammeter should not be used.

Errors.—There are a few special precautions to be observed in the use of thermal ammeters for large high-frequency currents. The leads should be brought straight in to the instrument; if run alongside the instrument so as to be parallel to its working parts, the current in the lead will have a greater mutual inductance with some of the current paths than others and will cause a change in current distribution which may disturb the reading. In using any ammeter of the hot-strip type care must be taken as to the position of the instrument. The readings vary with position because of altered heat convection from the strip. This effect is approximately the same at low and high frequencies. Error is avoided by always using the instrument in the position in which it was calibrated. The difference between the readings with the instrument in such positions that the strip is horizontal and strip vertical has been found to be of the order of 10 per cent.

All thermal instruments show a zero shift, which is troublesome in accurate measurements. In the usual ammeters it may amount to several per cent of full-scale deflection. To make a measurement of any accuracy it is necessary to leave the current on for at least several seconds, because the wire does not heat up instantaneously and the pointer does not come immediately to its final position. When the current is then cut off, the pointer does not return exactly to its original position. The difficulty may be overcome when the instrument has a zero adjustment by the following procedure: Allow the current to flow for several seconds without making a reading, cut it off, and after a certain time, say n seconds, adjust the zero, allow current to flow n seconds, and read; cut off current, and. if after n seconds the pointer is not exactly on zero, repeat the operations. The exact number of seconds to be allowed for each operation depends on the particular instrument. If the instrument has no zero adjustment, an accurate measurement can only be made by repeating the procedure followed in the calibration of the instrument. If the calibration procedure is not known, the most accurate method of reading is probably to first allow the current to flow for n seconds without reading, cut off, and after n seconds read the zero position of pointer, allow current to flow n seconds and read, cut off, and again take a zero reading after n seconds. These same methods of overcoming zero error apply to all thermal ammeters. In instruments having a nonuniform scale the zero reading must not be made in scale divisions; it should be estimated in millimeters, and the correction at any point found by multiplying by the number of scale divisions per millimeter at that point.

42. CURRENT TRANSFORMERS

Inductance and Capacity Shunting.—The reason why it is difficult to measure large currents of high frequency accurately is that ordinary shunts can not be used. As has been pointed out, the current divides between an instrument and its shunt according to

$$\frac{I_1^2}{I_2^2} = \frac{R_2^2 + \omega_2 L_2^2}{R_1^2 + \omega_2 L_1^2},$$

and the current ratio varies with frequency unless $\frac{L_2}{L_1} = \frac{R_2}{R_1}$. One way out of the difficulty has been explained in connection with thermal ammeters, viz, to make R_1 and R_2 so large in comparison

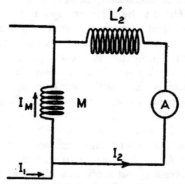

FIG. 104.—*Inductance shunted ammeter*

with L_1 and L_2 that the resistances determine the currents even for radio frequencies. This is done by use of wires of small diameter and high resistivity. Another possible solution is the opposite of this, viz, to make R_1 and R_2 negligible in comparison with L_1 and L_2 at radio frequencies. The method is, in short, to use inductance shunts. If any low-range ammeter of small resistance and inductance is connected in series with a large inductance L_2', the combination being shunted by a small inductance M, and if resistances are negligible, the current divides between the two paths according to

$$\frac{I_m}{I_2} = \frac{L_2'}{M}$$

It follows that

$$\frac{I_m + I_2}{I_2} = \frac{L_2' + M}{M}$$

Now $I_m + I_2 = I_1$, the total current.
Denoting the total inductance in the ammeter circuit by L_2, $L_2' + M = L_2$; then

$$\frac{I_1}{I_2} = \frac{L_2}{M} \qquad (74)$$

In practice the small inductance M might be a single loop of wire or even a length of straight wire, and the coil L_2' and the instrument A would be kept at such a distance from it that mutual inductance would have no effect. This method could not be used at low frequencies because the resistances would affect the current ratio.

A similar shunting scheme is the use of capacity shunts. A large condenser is inserted in the circuit. The capacity of this condenser (C in Fig. 105) is many times that of the main con-

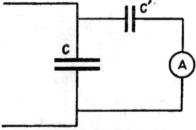

FIG. 105.—*Capacity shunted ammeter*

denser, with which it is in series. It is shunted by a small condenser C' in series with a low-range ammeter. The arrangement has the desirable feature of introducing only a small impedance into the circuit, because the capacity of condenser C is large. The resistances in ammeter and condensers must be kept small in order that the shunt ratio may be invariable. Care is necessary to prevent stray capacities, from the body of the observer, etc., from affecting this branch circuit. The posibilities of this method have not been studied experimentally.

Simple Theory of Current Transformer.—Inductance shunting is not practiced as shown in Fig. 104. The principle, however, is made use of in the current transformer. The instrument circuit is inductively coupled to the main circuit, as shown diagrammatically in Fig. 106, instead of direct coupled as in Fig. 104. If the resistance and inductance of the ammeter are negligible and there is no appreciable energy loss in the coils or the medium, equation (74) gives the ratio of currents. The resistance of the usual ammeters is of the order of several ohms and sometimes can not be neglected. The emf induced in the secondary circuit is $\omega M I_1$. This is opposed by the inductance L_2 and by the resistance R_2 of the ammeter.

$$\omega M I_1 = \sqrt{R_2{}^2 + \omega^2 L_2{}^2}\, I_2$$

$$\frac{I_1{}^2}{I_2{}^2} = \frac{R_2{}^2 + \omega^2 L_2{}^2}{\omega^2 M^2} = \frac{L_2{}^2}{M^2}\left(1 + \frac{R_2{}^2}{\omega^2 L_2{}^2}\right)$$

$$\frac{I_1}{I_2} = \frac{L_2}{M}\left(1 + \frac{R_2{}^2}{2\omega^2 L_2{}^2}\right), \tag{75}$$

(R_2 being small in comparison with ωL_2).

The current ratio is that given by equation (74), with a small correction term added. These calculations assume sine-wave currents, but apply to slightly damped currents as well. If the

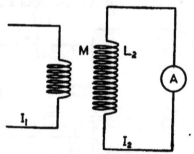

FIG. 106.—*Inductance shunted ammeter inductively coupled. Principle of the current transformer*

logarithmic decrement of the current is greater than a few hundredths, an additional correction term is needed. In using equation (75), the actual high-frequency value of L_2 must be used.

Transformer Without Iron.—The current transformers used in the measurement of radio currents are of two types, with and without an iron core. The iron core has advantages in certain circumstances; this is discussed below. The simple transformer without a magnetic core is, however, satisfactory if used carefully. A form which has been found[18] successful has a second-

<hr>

[18] See reference No. 103, Appendix 2.

ary winding consisting of a single layer of stranded wire on an insulating cylinder and a primary of one or more turns of thicker stranded wire near the middle of the secondary winding. To avoid induction from the leads and other parts of the circuit two such coils are used, connected so as to give astaticism, and the ends of the primary winding are brought out to a considerable distance. Equation (74) for the current ratio has been found to apply to such a transformer, within the accuracy of observation, for ratios as high as 100 to 1. The ratio may vary a few per cent with the frequency because of the resistance of the instrument connected to the secondary. At frequencies below the radio range the correction becomes very large.

Iron-Core Transformer.—The iron-core radio current transformer [18a] consists of a laminated iron ring with a close winding of one or a few layers of fine wire upon it and a small number of primary turns of heavy stranded wire linking with it, as shown in Figs. 107 and 109, facing page 157. Very thin silicon-iron sheet is a satisfactory core material. The ring may be very small, of the order of 5 cm diameter. Little care need be taken to avoid induction from other parts of the primary circuit, as in the use of the transformers without iron. The iron core greatly increases L_2 and insures close coupling between the primary and secondary turns.

The current ratio is readily found in terms of the ratio of primary and secondary turns. The self-inductance of the secondary winding is given by

$$L_2 = \frac{4 n_2^2 A \mu_a}{10^9 d}$$

where n_2 = number of secondary turns, A = area of cross section of iron, μ_a = apparent permeability of iron at the actual frequency used, and d = mean diameter of the iron ring. The mutual inductance is

$$M = \frac{4 n_1 n_2 A \mu_a}{10^9 d}$$

n_1 being number of primary turns. It follows that

$$\frac{L_2}{M} = \frac{n_2}{n_1}$$

[18a] See reference No. 103, Appendix 2.

Since the ratio of primary to secondary current was found above to be approximately $\frac{L}{M}$ it follows that in the iron-core transformer the current ratio is approximately the ratio of turns. The equation previously derived for current ratio does not apply exactly to the iron-core transformer because of the assumption of no energy loss. There is an energy loss in the iron due to eddy currents and hysteresis. This requires an energy current in the primary, which disturbs the current ratio. Taking account of this, the ratio may be shown to be

$$\frac{I_1}{I_2} = \frac{n_2}{n_1}\left(1 + \frac{aR_2}{\omega L_2}\right),\tag{76}$$

in which a is a quantity depending on the energy loss in the iron, having a value which is usually slightly less than unity. This assumes that all the magnetic flux from the secondary circuit links with the primary turns, a condition which is not fulfilled unless the secondary is uniformly and closely wound and the inductance of the instrument connected to the secondary is negligible.

Because of the iron core the secondary inductance L_2 is so large that the correction term in equation (76) is ordinarily negligible at radio frequencies. Thus an advantage of the iron-core transformer is that the current ratio does not vary appreciably with frequency and does not depend upon the instrument connected to it. Careful design is necessary to secure this constancy of ratio, and even then it holds only for radio frequencies. At low frequencies these current transformers have large errors. The reason is easily seen, since the correction term in equation (76) increases as ω decreases. The increase of the correction is not proportional to the decrease of ω because L_2 is smaller for high frequencies than for low. The value of L_2 is proportional to the apparent permeability of the iron, which decreases with increase of frequency because the skin effect reduces the effective cross section of the iron. Thus in a certain transformer [19] the iron had a permeability of 1000 at 50 cycles and an apparent permeability of 30 at 200 000 cycles. The correction term in the current ratio equation was 14 per cent at the lower frequency and 0.2 per cent at the higher.

---

[19] See reference No. 104, Appendix 2.

The apparent permeability depends on the thickness of the iron laminations. Making them thinner improves the accuracy of the transformer in two ways—it increases the apparent permeability and thus increases L_s, and it also makes a larger proportion of the magnetic flux from the secondary link with the primary turns. It is interesting to note that, at a given frequency, the apparent permeability of the iron does not vary with the current, because the fluxes in the iron are so very small that the permeability is practically constant.

Advantages of Transformer.—The ring form of current transformer has also been used without the iron core. Under certain conditions, as when an ammeter of extremely small resistance is connected to the secondary, this may be a very good form of instrument. Neither this nor the iron-core transformer has been exhaustively studied. Both forms, however, are definitely known to have the following advantages as devices for measuring large currents: (1) They conform to the requirement of simplicity of circuit, for the primary turns have very little inductance and capacity; (2) they utilize the magnetic effect of the current and are thus, in themselves, free from the inherent limitations of thermal ammeters such as thermal lag and dependence on surrounding conditions; (3) the measuring circuit is electrically insulated from the main circuit and there is thus no conducting path to the indicating instrument and so the capacity of the latter can not cause so great a loss of current from the main circuit.

Volt-ammeter Employing Current Transformer.—The current transformer is used in a portable measuring instrument designed at the Bureau of Standards for the use of the radio inspectors of the Bureau of Navigation, Department of Commerce. Two views of the instrument are shown in Figs. 108 and 109, facing page 157. This instrument, which is called a volt-ammeter, is a combination of a current-square meter, two high-frequency current transformers, and series resistances which may be thrown into the circuit when it is desired to use the instrument as a voltmeter.

The current-square meter is of a standard commercial type, requiring approximately 0.1 ampere for full scale deflection. Three scales are provided, running from 0 to 100, 0 to 5, and 0 to 25, respectively. The 0 to 100 scale is of use only in wave length or decrement measurement where relative values of current square are desired. The other two scales indicate amperes, and depend for their calibration on the accurate adjustment of the number of turns on the secondaries of the two current transformers contained in each instrument.

Views of the transformers are shown in Fig. 109. The cores of both transformers are composed of a number of ring-shaped laminations of very thin silicon steel. These are bound tightly together and served with a layer of empire cloth tape. Over this is uniformly wound a single layer of fine wire comprising the secondary. The number of secondary turns is determined from the relation

$$n_2 = n_1 \frac{I_1}{I_2} \qquad\qquad (77)$$

where

n_2 = number of secondary turns in series,
n_1 = number of primary turns in series,
I_2 = secondary current (about 0.1 ampere),
I_1 = primary current, either 5 or 25 amperes.

In these transformers the number of primary turns (n_1) adopted was 2, so that expression (77) becomes

$$n_2 = 2 \frac{I_1}{I_2}$$

This relation holds very closely at high frequencies if, as in this case of a toroid winding, the magnetic leakage is small or negligible. Small magnetic leakage requires that the secondary be wound in a single layer as uniformly as possible, covering the entire core length. The primary is wound in the manner indicated in Fig. 107, facing page ——, the wire being supported at a distance from the secondary and core. The terminals of the primaries of the transformers are brought out to four large binding posts at the bottom of the instrument. The secondary terminals are connected to opposite sides of a double-pole double-throw switch which is arranged to place the meter in series with either of the two windings.

In operation, when it is desired to measure the antenna current of a transmitting set, the meter is connected in series with the antenna and ground at the large binding posts marked 5 or 25 amperes, depending upon the magnitude of the current to be measured. The double-throw switch is then thrown to place the proper transformer in circuit and the readings obtained. The instrument will, of course, work equally well in a closed circuit.

FIG. 96.—*Hot-wire type current-square meter*

FIG. 102.—*Hot-strip ammeter with cylindrical arrangement of heating elements*

FIG. 111.—*Mounted thermocouple with protecting cap removed*

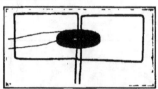

FIG. 107.—*Iron-core current transformer*

FIG. 108.—*Volt-ammeter employing the current transformer*

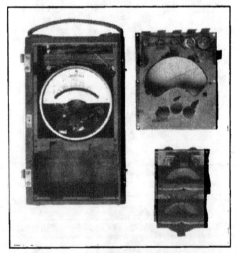

FIG. 109.—*Rear view of volt-ammeter, showing the current transformers*

As previously mentioned, the accuracy of the current scales depends upon the proper adjustment of the number of secondary turns. It is also affected by the frequency, the error growing larger as the frequency is decreased. These transformers are so constructed and calibrated that working over a range of frequencies corresponding to wave lengths between 150 and 1000 meters the current scales are accurate to better than 2 per cent.

The rubber-covered binding posts at the top of the meter, in conjunction with the push button marked "Voltage", are used for voltage measurement. Three ranges are provided, 2.7, 40, and 150 volts. These were adopted so as to enable the inspector to make proper voltage test on the storage batteries of the auxiliary power supply. Voltage calibration curves are supplied with the instrument.

The two sockets at the left marked decremeter connect directly to the terminals of the current-square meter. They are used when the meter is employed as a current indicator in wave length or decrement measurement.

This volt-ammeter was designed with the primary object of reducing the weight and number of instruments which the radio inspector must carry with him. In the wavemeter previously used by the inspectors the current-square meter was contained within the wavemeter case. In the latest type the current-square meter has been omitted, thus providing a much smaller and lighter wavemeter. This, together with the volt-ammeter here described, provides an equipment for making all the required measurements.

43. MEASUREMENT OF VERY SMALL CURRENTS

A number of methods are used for measuring currents of a few milliamperes or less. In addition to the thermal ammeter and the current transformer, used for larger currents, use is made of the electrostatic and the magnetic effects of the current, and rectification into unidirectional current. Instruments operating on these various principles are described below. The measurement of very small currents is free from some of the difficulties of measuring larger currents, principally because conductors of very small cross section may be used which do not change in resistance with frequency.

Crossed-Wire Thermoelement.—Sensitive thermoelements are easily made and are extensively used to measure small high-frequency currents. They consist essentially of two wires of different metals in contact, one or both of them being of very small diameter. The heat due to the $R I^2$ in the fine wire raises the temperature of the junction, thus giving rise to a thermoelectromotive force which is indicated by a direct-current galvanometer. A simple type is the crossed-wire thermoelement. Two fine wires are used, crossed in either of the ways shown in Fig. 110. (Practically the same type was shown in Fig. 98). The high-frequency current is led in through the heavy copper wires A and B, and the galvanometer is connected to a and b. The sensitivity of the thermoelement depends on the diameter, thermoelectric properties, and resistivity of the wires, on the length of the wires if very short, on the intimacy of contact of the two wires, and on the air pressure. Permanent contact of the two wires may be insured by soldering the junction. The use of a minute particle of solder does not appreciably reduce

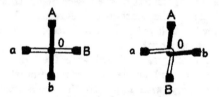

FIG. 110.—*Crossed-wire thermoelements*

the sensitivity. If the junction is not soldered, some thermoelements are occasionally found to have an abnormally high sensitivity, probably owing to a particularly poor contact; the high resistance of the contact causes a production of heat just at the junction which is large relatively to the heat produced in the wires. In some thermoelements this poor contact remains sufficiently constant so that the thermoelement can be relied upon, and in some it does not.

The materials ordinarily used for these thermoelements are constantan and steel, and constantan and manganin. Using wires of the order of 0.02 mm diameter and 4 mm long, a constantan-steel thermoelement has a resistance of about 1 ohm and gives about 40 microvolts for 15 milliamperes high-frequency current. Using a galvanometer with a sensitivity of 2.5 mm per microvolt, it follows that a deflection of 100 mm is produced by a high-frequency current of 15 milliamperes. The electromotive force is very closely proportional to the square of the high-frequency current. Thus, the typical thermoelement just mentioned gives a deflection of only 1 mm for 1.5 milliamperes.

These thermoelements have practically no thermal lag because of being made of such fine wire; it therefore pays to use a very quick-acting galvanometer with them. A thermoelement of this type made at the Bureau of Standards is shown in Fig. 111, facing page 156.

Smaller currents can be measured by the use of thermoelements made of wires of still smaller diameter. The resistance, the RI^2, and therefore the temperature, would be higher for a given current. High resistance is, however, objectionable in most radio circuits, so that the more sensitive thermoelements of higher resistance have little application.

Another way to increase sensitivity at the expense of increasing the resistance of the circuit is to connect the thermoelement, not directly into the circuit, but to the low-voltage side of a current transformer in the circuit. This has the advantage that the galvanometer is not metallically connected to the main circuit and thus its capacity is less likely to cause leakage of high-frequency current from the main circuit. The exact limitations of this method are not known; it seems likely that the calibration will change with frequency and with the resistance and inductance connected to the transformer.

The above figures on sensitivity are for thermoelements in air at ordinary atmospheric pressure. No variation of sensitivity with the ordinary barometric pressures has been observed, but the sensitivity may be greatly increased by placing the thermoelement in a vacuum. It has been found that an air pressure of about 0.01 mm of mercury or lower is necessary to gain much in sensitivity, but that in such low vacua the sensitivity of thermoelements of polished metal wires may be increased as much as 25 times. The removal of the air eliminates the cooling of the thermoelement by convection; a given current therefore raises it to a higher temperature. The temperature of a hot body in a vacuum is limited only by radiation of heat from its surface; thus the temperature of a polished metal surface, which is a poor radiator, rises higher than that of a dull metal surface, which is a good radiator.

Self-Heated Thermoelement.—In the type of thermoelement described in the preceding section the high-frequency current does not pass through the wires of the thermocouple itself. The thermocouple wires (*Oa* and *Ob*, Fig. 110) touch the wires (*OA* and *OB*), which carry the high-frequency current, at one

85601°—18——11

point only. There is no reason why the thermocouple wires themselves may not carry the high-frequency current and heat the junction. A simple thermocouple so used may be called a self-heated thermoelement. By using this type, Austin[20] has found it possible to utilize tellurium as one of the metals of a thermoelement and obtain very high sensitivity. A couple consisting of tellurium and platinum gives about 25 times as great an emf as constantan and platinum at the same temperature. These thermoelements are made by the following process.

Two copper wires are placed side by side about 3 mm apart and embedded in insulating material with their ends protruding (Fig. 112). To the end of one of these is soldered about 5 mm of 0.02 mm platinum or constantan wire. To the other is soldered a short bit of 0.8 mm platinum wire, to the end of which a bead of tellurium had previously been attached when the platinum was white hot. (White-hot platinum wire when inserted in

FIG. 112.—*Platinum-tellurium thermocouple of the self-heated type*

tellurium forms practically a resistance-free contact.) The end of the fine wire is next allowed to rest against the tellurium and the two are welded together electrically by means of a small induction coil with a high resistance in series with the secondary. The contact will be less fragile if the welding is done in an oxygen-free atmosphere. The resistance of a thermoelement prepared in this way may be anywhere from 5 to 50 ohms, according to the conditions of welding and the resistance of the fine wire, the lower resistance being somewhat more difficult to obtain. The thermoelement is next inclosed in a test tube, and, if likely to be handled roughly, the whole may be inclosed in a larger test tube with cotton or felt between the two. These thermoelements remain constant over considerable periods, but some have been found to lose their sensitiveness after a year or two. A 32-ohm thermoelement was found to give a deflection on a very sensitive galvanometer of 1 mm for 120 microamperes. Such thermoelements are very satisfactory when used in connection with portable microammeters of the pointer type.

Thermoelements of the self-heated type, can, of course, not be used on direct current, since the galvanometer forms a shunt on the heating wire. At high frequencies the current is kept out of the galvanometer because its impedance is so much greater than

[20] See reference No. 207, Appendix 2.

that of the short lengths of thermocouple wires which it shunts. It should be noted that the heating current passes through both junctions of the thermocouple but that the temperature of only one rises appreciably because a fine wire of high resistance is used at one junction only. Large cross section at the other junction prevents a large heat production and temperature rise.

Thermogalvanometer.—Currents of several hundred microamperes may be conveniently measured by the Duddell thermogalvanometer. This is a compact combination of hot wire, thermocouple, and galvanometer. The galvanometer coil is a single slender turn of wire, with the bismuth-antimony thermocouple attached to its lower end. The junction between these

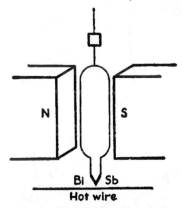

FIG. 113.—*Duddell thermogalvanometer*

two metals is directly over, but not in contact with, the heater which is either a hot wire or thin gold leaf or film of platinum on glass. Currents as low as 10 microamperes may be measured when a heater of several thousand ohms resistance is used, but this great sensitivity is not available for radio measurements, which generally require heaters of less than 50 ohms resistance. The thermogalvanometer differs from the thermoelements described above in that the thermocouple is not in metallic contact with the heater, and thus the capacity of the galvanometer is less likely to cause leakage of high-frequency current from the main circuit.

Bolometer.—An instrument in which heat is measured by the change of resistance which it produces in a conductor is called a bolometer. Precise measurements of radiant heat, for instance, are made with a bolometer which consists of a blackened metal strip together with a Wheatstone bridge for measuring its resistance. The bolometer used in radio measurements consists essentially of a fine wire, through which the high-frequency current is passed, connected to a Wheatstone bridge. The resistance of the wire increases as it is heated by the current, and this change

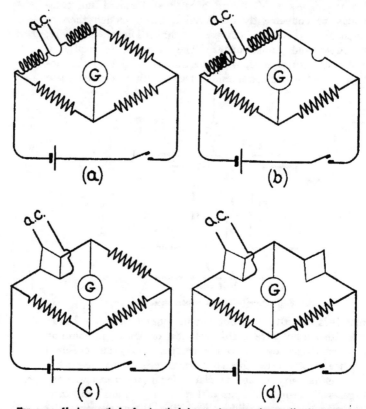

FIG. 114.—*Various methods of using the bolometer for measuring small radio currents*

of resistance either causes a deflection of the bridge galvanometer, or the bridge is balanced by a change in one of the bridge arms so as to keep the galvanometer deflection unchanged. In the latter method the resistance in the changed bridge arm is a measure of the current.

The direct current used in the Wheatstone bridge produces more or less heat in the bolometer wire, and in order to avoid error it is necessary to: (1) Keep it very small, or (2) keep it constant for all measurements, or (3) have an auxiliary wire similar to the bolometer wire in an adjacent arm, which will keep the bridge balanced as far as heating by the direct current in the bridge is concerned. The various forms shown in Fig. 114 have been used. Figures (b) and (d) are similar to (a) and (c) respectively, except for the auxiliary wires in a bridge arm adjacent to the one containing the bolometer wire. It is necessary to keep the high-frequency current out of the parts of the bridge other than the bolometer wire as the latter would not then carry the whole current to be measured. This is done in either of the two ways shown. Choke coils may be used on either side of the hot wire, as in (a) and (b); or, the bolometer wire may have the rhombus form as in (c) and (d). The heating current divides between the two halves of the rhombus, and the bridge connections are made at two points of equal potential so that there is no tendency for the heating current to flow into the bridge.

For the bolometer wire, use has been made of iron, gold, platinum, tungsten, and carbon. The smaller the wire the smaller the currents measurable. A gold wire 0.002 mm diameter has been found to give 10 scale divisions on a pointer-type galvanometer for 500 microamperes and a 0.0005 mm platinum wire 10 scale divisions on the same galvanometer for 34 microamperes. These figures [21] are for the bolometer wire in air. The sensitivity may be increased by placing in a vacuum. A current of 5 microamperes has been measured by the use of a carbon filament in a vacuum.

The bolometer has also been used as a means of measuring large currents up to 10 amperes. (See p. 172 below.) Wire of comparatively large diameter is used but not so large as to change in resistance with frequency. It is immersed in oil to keep down the the temperature rise.

Electrometer.—It is possible to measure fairly small currents by the aid of an electrometer shunted across a condenser. The deflections of an electrometer are proportional to the square of the effective voltage when the vane is connected to one plate. The current through the electrometer is proportional to its capacity, to the frequency, and to the applied voltage. The form shown in Fig. 116 has been used. Connection is made to the fixed plates *PP*. The light metal vane *V* is suspended by a delicate fiber between them. The suspending fiber also

[21] Data from B. Gáti (Electrician, 58, p. 983, 1907, and 78, p. 354, 1916), who uses an arrangement for measuring resistance somewhat different from a Wheatstone bridge, and who calls his device a "barretter."

carries a damping-vane and a mirror for reading from a distance.
The voltage impressed on the plates PP charges them with electricity of opposite signs.   One of them being connected to the vane
V, they thus exert a torque on it.

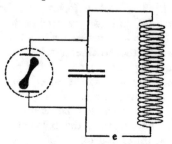

FIG. 115.—*Method of using electrometer (vane
should be connected to one terminal)*

These instruments have the advantage of introducing no appreciable resistance into the circuit.   They have a small capacity and
therefore should not be used in parallel with any except a large
condenser.   They require very careful manipulation, as they are
delicate and are very sensi-
tive to stray electrostatic
charges.   The method has
not been used in the Bureau
of Standards laboratory.

FIG. 116.—*Schematic ar-
rangement of plates
and vane of the electro-
meter*

Electrodynamometer.—The
instruments of this type used
at low frequency consist of a
fixed and moving coil connected either in series
or in parallel.   Capacity between the two coils
renders them unsuitable for measuring high-
frequency currents.   A somewhat different type
has, however, been successfully used.   The instrument consists of a small coil of fine wire
which carries the current to be measured, and
a flat ring or disk of silver or copper suspended
by a delicate fiber concentric with the coil and
with its plane at 45° to the plane of the coil.

FIG. 117.—*Type of elec-
trodynamometer suit-
able for radio current
measurement*

Current in the coil induces an opposing current in the ring, which
is then repelled.   The torque acting on the ring is proportional to

$$\frac{LN\omega^2 I^2}{R^2 + \omega^2 L^2},$$

where R and L are the resistance and inductance of the metal ring and N the number of turns in the coil. The deflection will therefore depend on the frequency, which is a serious limitation on the usefulness of the instrument. For frequencies so high, however, that R is small compared with ωL, the deflections are independent of frequency. Care should be taken to keep the instrument away from other parts of the circuit, so that no large magnetic fields may act upon it.

Crystal Detector.—Currents too small to be measured by any of the preceding instruments, as for example the received currents in antennas, can be measured by the aid of a crystal detector.

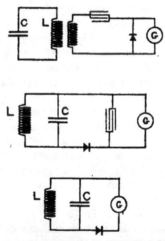

FIG. 118.—*Crystal detector circuits used in measuring small currents*

The exact action of these detectors is a complicated matter, but for practical purposes it is sufficient to regard them as unilateral conductors; that is, they have a greater resistance to current flowing through them in one direction than to current flowing in the opposite direction. Thus, when an alternating emf is impressed on a crystal detector, more current flows in one direction than in the other, and a direct-current instrument in the circuit will be operated. The resistance of the ordinary crystal detectors in the low-resistance direction is of the order of 1000 to 10 000 ohms, and the resistance in the opposite direction about 10 times as great.

In combination with a sensitive galvanometer, a crystal detector may be used to measure currents of a few microamperes. The galvanometer and crystal may be connected to the circuit *LC*, the current in which is to be measured, in any of the ways shown in Fig. 118. Using the first mode of connection, Austin [22], has obtained a deflection of 100 mm for 91 microamperes. A perikon detector (chalcopyrite-zincite) was used, with a galvanometer of 2000 ohms resistance giving 1 mm for 1.3×10^{-9} ampere direct current. The deflections have been found to be proportional to the square of the high-frequency current (for this and some other crystals). The sensitiveness of the crystal detector is hundreds of times that of the thermoelement, but it is not constant. It is always calibrated just before or after use (or both) by comparison with a thermoelement in the *LC* circuit, using current from a buzzer or other source. (See p. 174 below.)

A telephone may be used in place of the galvanometer in any of the arrangements shown in Fig. 118, when periodically interrupted current is to be measured. Telephone measurements can not be made of uninterrupted undamped currents. Quantitative measurements may be made with the telephone in two ways. In both, the current through the telephone is reduced until the sound is just barely audible. (The limit of audibility is sometimes taken to be that at which dots and dashes can just barely be distinguished.) In the first method a resistance is placed in parallel with the telephone and reduced until the limit of audibility is reached; this is the "shunted telephone" method. The second method employs variable coupling between the detector circuit and the main circuit.

A measure of current in the shunted telephone method is obtained as follows: If *t* is the impedance of the telephone for the frequency and wave form of the current impulses through it, *s* the impedance of the shunt, I_t the least current in the telephone which gives an audible sound, and *I* the total current flowing in the combination of telephone and shunt,

$$\frac{I}{I_t} = \frac{s+t}{s}$$

This ratio, $\frac{s+t}{s}$, is called the audibility. It is approximately proportional to the square of the high-frequency current. It can be expressed in units of current if calibrated at one or more values of current by some deflection device in the high-frequency circuit.

---

[22] See reference No. 206, Appendix 2.

The observed settings depend on the frequency, on the wave form of the pulses passing through the telephone, on the constants of the circuit, and on the frequency of interruption of the current used, and involve the assumption that the crystal and the sensitiveness of the operator's ear remain constant. It is desirable to minimize the variations of resistance which have to be made in the detector circuit by using a fixed resistance R_1 in series with the detector and shunt the telephone across a variable portion R_2 of this. The apparatus should be calibrated under the exact conditions of use, both before and after each set of measurements.

An accuracy of 10 per cent is difficult to obtain; the method is nevertheless very useful in measurements of radio currents in receiving sets. For a frequency of interruption of the radio current of 1000 per second, currents of the same order of magnitude can

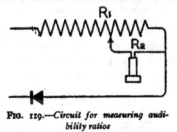

FIG. 119.—*Circuit for measuring audibility ratios*

be measured with a crystal detector by the use of a telephone as by the use of a galvanometer. About 10 microamperes is the least current which can be detected by the ordinary crystal and telephone.

In the variable coupling method the telephone is not shunted, but the coupling between the detector circuit and the main circuit is varied until the sound in the telephone is just barely audible. The greater the high-frequency current to be measured the looser is the coupling. The arrangement is calibrated by making at least one simultaneous observation of the coupling for barely audible sound and current as measured by some other device, such as a previously calibrated crystal and galvanometer connected to the main circuit, together with the plotting of a curve between coupling and current in the main circuit. The coupling may be measured in any arbitrary way, as by the distance apart of the coupling coils.

Audion.—The audion (described below in section 56) may be used for measurements of current, just as the crystal detector is, in conjunction with either a telephone or a galvanometer. With the ordinary audion connections, as in Fig. 120, the sensitivity is about the same as that of the best crystal detectors. The actions of the audion and other electron tubes as detectors, amplifiers, etc.,

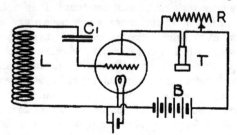

FIG. 120.—*Use of the audion for measuring small currents in terms of audibility ratios*

are discussed on pages 204 to 210. The connections shown here are for the shunted telephone method. The variable coupling method can also be used. In the figure, L is the coil used for coupling to the circuit in which the current is to be measured, C_1 is a small

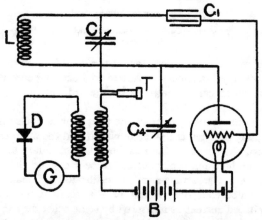

FIG. 121.—*Oscillating ultraudion circuit used with crystal detector and galvanometer for measuring small currents*

fixed condenser, and T the telephone shunted by the variable R. The audibility is approximately proportional to the square of the high-frequency current, as in the case of the crystal detector.

A galvanometer can be used with the audion, but it must not be put directly in place of the shunted telephone, because a continuous current would flow through it from the *B* battery. One arrangement is to place the primary of a transformer in series with the telephone and connect to its secondary a crystal detector in series with a sensitive galvanometer. The changes in current which affect the telephone give rise to alternating currents in the secondary which are rectified by the crystal detector and thus cause a deflection of the direct-current galvanometer. This arrangement is particularly advantageous when the oscillating ultraudion connections are used. (For description of the ultraudion see section 58 below.) The connections are given schematically in Fig. 121. This is suitable for the measurement of undamped currents. The note in the telephone *T* is produced

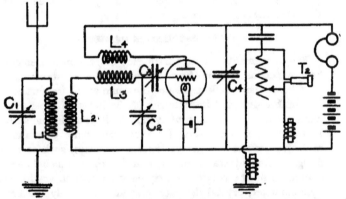

FIG. 122.—*Oscillating audion circuits for quantitative measurements on undamped waves from distant radio stations*

by the beats between the impressed and the local currents. The condenser C_4 must be adjusted for maximum deflection of the galvanometer. Austin[28] has found that the deflections are proportional to the square of the high-frequency current, which means that the current in the telephone is proportional to the first power of the high-frequency current. (This law holds only for the oscillating condition. When the audion is not oscillating, the deflections are approximately proportional to the fourth power of the high-frequency current.) This constitutes a method of remarkable sensitiveness for measuring small high-frequency currents. Austin found that for signals of minimum audibility on the simple audion, the oscillating ultraudion gave audibilities from 300 to 1000 times as great; that is, it would measure currents hundreds of times as small.

For convenience in measuring received radio currents from distant stations the shunted telephone is used in connection with the oscillating ultraudion. The arrangement shown in Fig. 122 has been used by Austin.[24] The shunt s is used on the telephone T_2. The audibility is approximately proportional to the current in the antenna. The sensitivity is always measured at the time of use by comparison with a silicon detector and galvanometer, which combination is in turn calibrated by comparison with a thermoelement. This arrangement has been used to make quantitative measurements on undamped waves from radio stations 4000 miles away, the least high-frequency current detectable in the receiving antenna being 4×10^{-9} ampere.

44. STANDARDIZATION OF AMMETERS

The instruments for high-frequency current measurement may be grouped as follows, from the standpoint of standardization: (1) Those whose deflections are the same at all frequencies, such as suitably designed instruments of the hot-wire type; (2) those whose deflections are accurately calculable at all frequencies, such as electrometer ammeters; (3) those which are constant at all radio frequencies but not at lower frequencies, such as properly designed current transformers; and (4) those which have to be calibrated at the particular frequency used, such as the electrodynamometer, crystal detector, and audion. Only the first and second of these groups are suitable to serve as standards for the calibration of high-frequency ammeters. The second group, electrometers, is not actually used for this purpose, so the ultimate standards used in practice are instruments of the hot-wire type.

Small and Moderate Currents.—The instruments described in subsection a above under this head are all simple hot-wire types. If properly constructed, if the hot wire is fine enough, and the design is otherwise correct, they are themselves standards and need no calibration at high frequency. Such instruments are calibrated at low frequency (50 to 3000 cycles per second). In no case should they be used without calibration. While thermocouples and some detectors give deflections approximately proportional to the square of the current, they do not follow this law

---

[24] See article in "Electrician," reference No. 108, Appendix s.

exactly. In some cases the calibration can be made with direct current. In other cases this is not desirable; in the thermoelectric ammeter of the crossed-wire type, for instance, the Peltier effect at the junction causes a current through the galvanometer, and in addition some of the direct current passes through the galvanometer inasmuch as the junction has some resistance and thus the galvanometer and the junction constitute two parallel paths for the current. Both of these effects may be eliminated by reversing the direct current and taking the mean deflection of the galvanometer (not the mean reading, in case the scale is calibrated in terms of the heating current). Reversing the direct current is equivalent to using alternating current for calibration.

The use of direct current involves another possible error, leakage to the galvanometer, which may or may not be reversed when the current is reversed. It is on the whole good practice to use alternating current rather than direct for standardizing high-frequency ammeters, using as comparison instruments any reliable low-frequency ammeters.

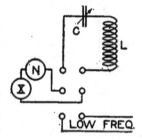

If there is any doubt as to a high-frequency ammeter's independence of frequency, it should be compared with a reliable standard at radio frequencies by the methods given below.

FIG. 123.—*Method of testing ammeters for the effect of change of frequency*

Large Currents.—The design of most ammeters for large currents of high frequency is such that it is not safe to assume them independent of frequency. They should be standardized by comparison with instruments known to be reliable at several radio frequencies. The comparison is made as indicated in Fig. 123. The instrument to be tested, X, is in series with a standard instrument, N, and their deflections are simultaneously observed when supplied alternately with high-frequency and low-frequency current. The high-frequency circuit LC is coupled to a source of current such as a spark or arc set or a pliotron, and the low-frequency current is obtained from an alternator through a step-down transformer and a rheostat.

The two ammeters could, of course, be compared using the high-frequency current only, but this would give no information as to the change of reading from low to high frequency. Also, the variation of the readings at different radio frequencies would

not be as accurately obtained. The particular advantage of
using an auxiliary low-frequency comparison current is that it
enables one to determine accurately the difference between the
high and low-frequency readings independently of zero shift,
temperature variation, and other accidental errors. The experimental procedure is to pass high-frequency current through the
two ammeters for a certain length of time, say, one minute,
recording the deflections, then quickly throw the switch (*S* in
Fig. 123) and allow an approximately equal low-frequency current to flow the same length of time, recording the deflections;
then high frequency again, then low frequency, and finally high
frequency again. Thus, three high-frequency observations are
obtained with two low-frequency observations sandwiched between
them, and errors from thermal or other drifts are eliminated.
The lack of constancy of a spark or arc source limits the precision

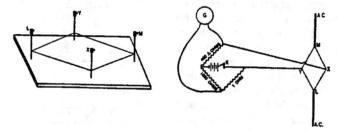

FIG. 124.—*Standard ammeter using the bolometer principle*

of an observation to a few tenths per cent. This can be excelled
with a pliotron source.

Several variations of the hot-wire principle are available as
standard ammeters. The simplest is a single wire as used in
various instruments previously described, and capable of measuring
up to about 2 amperes. An ammeter of about 10 amperes range
can be calibrated by comparison with such a standard at the lower
end of its range, 2 amperes or less. Such a calibration is, of course,
not so satisfactory as one covering the whole range.

As a standard ammeter for measurements up to 10 amperes
an application of the bolometer principle may be used. The
instrument consists essentially of a fine copper wire soldered
to four upright posts, two of which carry the high-frequency
current and the other two connect to a Wheatstone bridge as
in Fig. 124. The wire rhombus is placed in oil. The high-
frequency current has two paths in the instrument and hence

great care is necessary to insure that the resistances and inductances of the two paths are equal. With the most careful construction the current distribution between the two paths doubtless
varies somewhat with frequency, but it is to be noted that the
resistance depends upon the heat production in the whole instrument and not on that in one branch only. Small changes of
current distribution do not appreciably affect the resistance of the
instrument, and it is consequently a perfectly reliable high-
frequency standard if carefully constructed.

A curve is plotted between the measured current and the
resistance in the rheostat arm of the Wheatstone bridge. In
the arrangement shown diagrammatically in Fig. 124, K is a
tapping key in the battery circuit. A closed galvanometer
circuit is used, thus eliminating errors of false zero. G is a sensitive moving-coil galvanometer. The current through the
fine wire from the bridge battery should be 0.01 of the heating
current or less. It is not convenient to calibrate this standard
on direct current, although it is theoretically possible to do so.
For a heating current entering at L and M, X and Y need to be so
adjusted as to be equipotential points; then no portion of the
heating current flows in the bridge used to measure the resistance.
However, it is difficult to make this adjustment exactly, and it is
moreover unnecessary, as a calibration by low-frequency alternating current is just as good as a direct-current calibration.
Consequently, the points X and Y are simply made approximately
equipotential points, but not adjustable.

For currents greater than 10 amperes, standard ammeters
of the cylindrical type, described on page 146 above, are used. As
there stated, the design of these instruments by no means insures
their accuracy. If properly constructed of high-resistance metal,
instruments of this type can be used as standards for moderate
ranges of current and frequency. Standard instruments for the
largest currents used in radio work are now under investigation
at the Bureau of Standards.

Very Small Currents.—Thermoelements are usually used as
standards in calibrating the instruments used to measure very
small currents. As already explained, it is in general better to
standardize them with low-frequency alternating current than
with direct current. In particular, thermoelements of the self-
heated type can not be used on direct current. It is possible to
use a bolometer as a standard, and it likewise should be calibrated
with low-frequency alternating current, since if direct current

were used some of it would be very likely to get into the bridge galvanometer. The thermogalvanometer and the electrometer, on the other hand, may theoretically be calibrated with direct current but the experimental errors are likely to be larger than with low-frequency alternating current.

In the use of the crystal detector and the audion to measure current, with either a galvanometer or telephone, it is necessary to calibrate the arrangement at the time of use. These very sensitive devices are variable with time and with the conditions of the circuits. The calibration of a crystal detector is made by placing a thermoelement directly in the circuit to which the crystal is coupled or attached, as in Fig. 125, and currents used

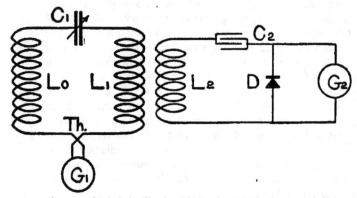

Fig. 125.—*Method of calibrating a detector in terms of a thermocouple*

such as to give small deflections on the thermoelement galvanometer. These observations fix the value of current for one or more points corresponding to large deflections of the galvanometer attached to the crystal. The law of variation of deflection of the latter galvanometer with respect to current in the main circuit is determined by a separate experiment in which the detector is coupled more loosely to the main circuit so that the deflections of the two galvanometers are more nearly equal. In Fig. 125, L_0 is an inductance used to couple the main circuit to the source of current, Th is the thermoelement, and G_1 its galvanometer. L_2 is used to couple the detector circuit to L_1, C_2 is a fixed condenser, and G_2 is the detector galvanometer.

Ordinarily the deflections of the two galvanometers are approximately proportional, and the ratio between the small thermoelement deflection and the large crystal deflection obtained in the calibration is used as a multiplier to obtain the equivalent thermo-element deflection from any smaller crystal deflection subsequently observed. The currents so measured thus depend, in a certain sense, on an extrapolation. Great accuracy is not expected nor required. When an oscillating ultraudion is used it is calibrated by comparison with a detector in the same way as the detector is calibrated in terms of the thermoelement. The exceedingly small currents measured with the oscillating ultra-udion thus depend upon two extrapolations.

RESISTANCE MEASUREMENT

45. HIGH-FREQUENCY RESISTANCE STANDARDS

Standards of resistance are required for some of the methods of resistance measurement described below. The resistance of the standard must be accurately known at all frequencies, and it is very desirable to have it remain constant over all the frequencies at which it may be used. This requirement practically limits the form of such a standard to a very fine wire. Very thin tubes would be satisfactory from the theoretical standpoint, but it is extremely difficult to obtain very thin metal tubes of sufficient uniformity that the current distribution, and hence the resistance, does not change with frequency.

The accuracy of ordinary measurements requires that the resistance of the standard be constant to 1 per cent. The maximum size of wire of various materials that may be used can be found from Table 18, page 310, for the highest frequency that is to be used. The diameter required for any given accuracy and a given limiting frequency may be calculated from (207), page 300.

These fine wires which must be used as high-frequency standards will not carry much current without serious heating. They must therefore be used in measurements with caution. When a standard is required to carry large currents it can not, in general, be obtained by putting several fine wires in parallel. The design of such a resistance standard is, in fact, identically the same problem as the design of high-frequency ammeters for large currents. (Sec. 41, p. 144.)

The ideal resistance standard would not change any of the constants of a circuit except resistance when it is inserted in the circuits. A wire, however, has some inductance, and since the inductances used in radio circuits are small the inductance of a wire standard of resistance can not be neglected. The inductance must be made extremely small by using very short wires, or its effect must be minimized by substituting for the resistance wire a copper wire of the same length whenever the resistance wire [is removed from the circuit. The first of these alternatives is followed in the use of a short slide wire, which gives a continuous variation of resistance; the contact with such a slide wire must, of course, be of small and constant resistance. The second alternative is followed in sets of resistance standards. The inductance of the copper wire link will be practically the same as the inductance of the resistance standard substituted for it, and the slight difference between them will not ordinarily affect the total inductance of the circuit provided the resistance wire and the copper link are short.

A set of resistance standards for high frequency may, therefore, be a set of short wires of approximately equal lengths, a portion of the length consisting of a very fine wire. The wire must be of very small diameter to obtain fairly large resistances in the short length allowed. This fine wire is of resistance material, the length is adjusted to give the required resistance, and the remainder of the length is of relatively thick copper wire. In a set used by this Bureau, the resistance material is manganin, used because the resistance does not change appreciably with temperature. In order to protect the delicate wires from breakage each is mounted in a glass tube. The copper ends of each standard are amalgamated for insertion in small mercury cups; the amalgamation must be renewed at frequent intervals. They are 7 cm long; the resistances have values from 0.2 to 40 ohms; the resistance wires have lengths from 0.5 to 6 cm, and diameters from 0.03 to 0.12 mm. On account of the wires being of such small diameter, care is necessary to avoid using currents that would overheat them. The inductances vary from 0.15 microhenry for the 40-ohm standard to 0.08 microhenry for the 0-ohm copper link. The difference, 0.07 microhenry, is negligible except in rare cases. The resistances have been found to remain satisfactorily constant for several years.

Decade resistance boxes are also useful as standards when properly made and used with caution. The resistance units must be made very short and of sufficiently fine wire. When such a set is used in a circuit of low inductance, variation of the resistance setting may vary the inductance of the circuit. In some methods of resistance measurement this merely requires retuning to resonance.

46. METHODS OF MEASUREMENT

The methods of measuring high-frequency resistance may be roughly classed as: (1) Calorimeter method, (2) substitution method, (3) resistance-variation method, and (4) reactance-variation method.

The fourth has frequently been called the "decrement method," but it is primarily a method of measuring resistance rather than decrement, exactly as the resistance-variation method is. Thus, the measurement of decrement is the same problem as the measurement of resistance. When applied to determine the decrement of trains of waves, radio-resistance measurement accomplishes something similar at high frequencies to what is done at low frequencies by wave analysis. Either may be used to measure the decrement of a wave under certain conditions and, in fact, the results of resistance measurement by any method may be expressed in terms of decrement.

All four methods may be used with either damped or undamped waves, though in some of them the calculations are different in the two cases. They are all deflection methods, in the sense of depending upon the deflections of some form of high-frequency ammeter. In the first and second, however, it is only necessary to adjust two deflections to approximate equality, while in the third and fourth the deflections may have any magnitude.

47. CALORIMETER METHOD

This method may be used to measure the resistance either of a part or the whole of a circuit. The circuit or coil or other apparatus whose resistance is desired, is placed in some form of calorimeter, which may be a simple air chamber, an oil bath, etc. The current is measured by an accurate high-frequency ammeter, and the resistance R_x is calculated from the observed current I

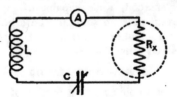

Fig. 126.—*Method of measuring resistance by calorimeter method*

and the power, or rate of heat production, P.

$$P = R_x I^2 \qquad (77)$$

While P might be measured calorimetrically, in practice it is always measured electrically by an auxiliary observation in terms of low-frequency or direct current. Thus, it is only necessary to observe the temperature of the calorimeter in any arbitrary units when the high-frequency current flows after the temperature has reached a final steady state, and then cause low-frequency current to flow in the circuit, adjusting its value until the final temperature becomes the same as before. Denoting by the subscript o the low-frequency values

$$P_o = R_o I_o^2$$

$$\frac{P}{P_o} = \frac{R_x I^2}{R_o I_o^2}$$

$$\frac{R_x}{R_o} = \frac{P I_o^2}{P_o I^2}$$

For $P = P_o$, $$R_x = \frac{R_o I_o^2}{I^2} \tag{78}$$

From the known low-frequency value of the resistance, therefore, and the observed currents, the resistance is obtained.

The high and low frequency observations are sometimes made simultaneously, using a duplicate of the apparatus whose resistance is desired, placed in another calorimeter as nearly identical with the first as possible. High-frequency current is passed through one, low frequency through the other, and the calorimeters kept at equal temperatures by means of some such device as a differential air thermometer or differential thermoelement. To compensate for inequalities in the two sets of apparatus, the high and low frequency currents are interchanged. This method may be found more convenient in some circumstances, but the extra complication of apparatus is usually not worth while, and the value of the measurement depends upon the accurate observation of the high-frequency current I, just as the simpler method does.

The calorimeter method, while capable of high accuracy, is slow and less convenient than some of the other methods. It has been used by a number of experimenters to measure the resistance of wires and coils.

48. SUBSTITUTION METHOD

This method is applicable only to a portion of a circuit. Suppose that in Fig. 126 the coil L is loosely coupled to a source of oscillations. The capacity C is varied until resonance is obtained, and the current in the ammeter is read. A resistance standard is then substituted for the apparatus R_x and varied until the same current is indicated at resonance. If the substitution has changed the total inductance or capacity of the circuit, the returning to resonance introduces no error when undamped or slightly damped electromotive force is supplied, provided the change of condenser setting introduces either a negligible or known resistance change. In the case of a rather highly damped source, however, the method can only be used when the resistance substitution does not change the inductance or capacity of the circuit. The unknown R_x is equal to the standard resistance inserted, *provided* the electromotive force acting in the circuit has not been changed by the substitution of the standard for R_x; this condition is discussed below.

Unequal Deflections.—The resistance standards usually used are not continuously variable, and hence the standard used may give a deflection of the ammeter somewhat different from original deflection.

To determine the resistance in this case, three deflections are required, all at resonance. The apparatus, say an absorbing condenser, of unknown resistance R_x is inserted and the current I_x observed; then a similar apparatus of known resistance R_N is substituted for it and the current I_N observed; and finally a known resistance R_1 is added and the current I_1 observed. The relations between these quantities and the electromotive force involve the unknown but constant resistance of the remainder of the circuit R, thus, assuming undamped emf,

$$R_x + R = \frac{E}{I_x}$$

$$R_N + R = \frac{E}{I_N}$$

$$R_1 + R_N + R = \frac{E}{I_1}$$

from which, $R_x - R_N = R_1 \dfrac{\dfrac{I_N}{I_x} - 1}{\dfrac{I_N}{I_1} - 1}$ (79)

Or, in case the resistance of a current-square meter, thermoelement, or similar apparatus is desired, the procedure would be to observe the current I when the resistance R is that of the circuit alone, the current I_x when the apparatus of resistance R_x is inserted in the circuit, and the current I_1 when a known resistance R_1 is substituted.

Then

$$R = \frac{E}{I}$$

$$R_x + R = \frac{E}{I_x}$$

$$R_1 + R = \frac{E}{I_1}$$

From which,

$$R_x = R_1 \frac{\left(\dfrac{I}{I_x} - 1\right)}{\left(\dfrac{I}{I_1} - 1\right)}$$

This method is closely related to the resistance-variation method; see formula (80) below.

Application of Method.—The substitution method is very convenient and rapid, and is suitable for measurements upon antennas, spark gaps, etc., and for rough measurements of resistances of condensers and coils. In radio laboratory work, however, using delicate instruments and with loose coupling to the source of oscillations, it is found that in some cases it is not a highly accurate method, except for measuring small changes in resistance of a circuit. The reason for this is that there are other electromotive forces acting in the circuit than that purposely introduced by the coupling coil, viz, emf's electrostatically induced between various parts of the circuit. When the apparatus under measurement is removed from the circuit, these emf's are changed, and there is no certainty that when the current is made the same the resistance has its former value. Something of the same difficulty enters into the question of grounding the circuit in the following method, as discussed below.

49. RESISTANCE VARIATION METHOD

As this method measures primarily the resistance of the whole circuit, the principle may be readily understood from the diagram of the simple circuit, Fig. 127.

If the resistance of some particular piece of apparatus, inserted at P, for example, is to be found, the resistance of the circuit is measured with it in circuit and then remeasured in the same way with it removed or replaced by a similar apparatus of known resistance; and the resistance of the apparatus is obtained by simple subtraction.

The measurement is made by observing the current I in the ammeter A when the resistance R_1 has its zero or minimum value, then inserting some resistance R_1 and observing the current I_1. Let R denote the resistance of the circuit without added resistance, and suppose a sine-wave electromotive force E introduced into the circuit by induction in the coil L from a source of undamped waves. For the condition of resonance

$$I = \frac{E}{R}$$

$$I_1 = \frac{E}{R + R_1}$$

from which the resistance of the circuit is given by

$$R = \frac{R_1}{\dfrac{I}{I_1} - 1} \tag{80}$$

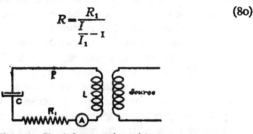

Fig 127.—*Circuit for measuring resistance by the resistance variation method*

The same method can be employed using damped instead of continuous waves, and can even be used when the current is supplied by impulse excitation, but the equations are different; see (89) and (90) below.

Precautions.—A limitation on the accuracy of the measurement is the existence of the emfs electrostatically induced that were mentioned above. In the deduction of (80) it is assumed that E remains constant. The virtue of this method is that these emfs may be kept substantially constant during the measurement of resistance of the circuit. They will invariably be altered by the substitution of the apparatus whose resistance is desired, but the resistance of the circuit is measured accurately in the two cases and the difference of the two measurements gives the resistance sought. In order to keep these stray electromotive forces unchanged when R_1 is in and when it is out of circuit, particular attention must be paid to the grounding of the circuit. The shield of the condenser and the ammeter (particularly if it is a thermocouple with galvanometer) have considerable capacity to ground and are near ground potential. A ground wire, if used, must be connected either to the condenser shield or to one side of the ammeter. If connected to the high-potential side of the inductance soil absurd results will be obtained. The resistance R_1 also must be inserted at a place of low potential, preferably between the shielded side of the condenser and ammeter.

Furthermore, care must be taken that the coupling between the measuring circuit and the source is not too close. Otherwise the current in the source and hence the emf E will vary somewhat when R_1 is inserted. This will give incorrect resistance values which will depend upon the magnitude of R_1. Whether such an effect is present can be judged by opening and closing the measuring circuit and noting whether this produces a considerable change in the ammeter reading in the source circuit; or by repeating the measurement with reduced coupling. In order that the measurement can be made using very loose coupling it is necessary either to have a source of considerable power or to use a sensitive current-measuring device such as thermocouple and galvanometer.

Use of Thermocouple.—As regularly carried out at the Bureau of Standards, in the resistance-variation method a pliotron is used as a source of undamped emf, and current is measured with a low-resistance thermocouple in series in the circuit. The currents corresponding to given deflections of the thermocouple galvanometer are obtained from a calibration curve; or from the law $d \propto I^2$, where d = deflection, if the instrument follows this law sufficiently closely. When the deflections follow this law, equation (80) becomes

$$R = \frac{R_1}{\sqrt{\dfrac{d}{d_1} - 1}} \tag{81}$$

Several values of resistance R_1 are usually inserted in the circuit and the corresponding deflections obtained; the resulting values of R are averaged. An example is given on page 192 below.

When the thermocouple follows the square law accurately, the quarter deflection method may be used, which eliminates all calculation. When the deflection d_1 is ¼ d, equation (81) becomes

$$R = R_1 \tag{82}$$

This method requires a variable resistance standard such that R_1 can be varied continuously in order to make d_1 just equal to

$\frac{1}{4}$ d. Practically the same method is used if the resistance is varied by small steps, as in a resistance box, and interpolating between two settings of R_1. Suppose that the two values of R_1 are R_2 and R_3 and the corresponding deflections d_2 slightly less than $\frac{1}{4}$ d, and d_3 slightly greater than $\frac{1}{4}$ d,

$$R = R_2 + \frac{d_2 - \frac{1}{4}d}{d_3 - d_2}(R_3 - R_2) \tag{83}$$

Use of Impulse Excitation.—The procedure for the resistance-variation method is the same when the current is damped as when undamped. When the circuit is supplied by impulse excitation, so that free oscillations are produced, the theory of the measurement is very simple. The current being I when the resistance is R and I_1 when the resistance R_1 is added, the power dissipated in the circuit must be the same in the two cases, because the condenser in the circuit is charged to the same voltage by each impulse which is impressed upon it, and there is assumed to be no current in the primary after each impulse.

Therefore, $RI^2 = (R + R_1)I_1^2$
whence,

$$R = R_1 \frac{I_1^2}{I^2 - I_1^2} \tag{84}$$

This simple deduction of this equation is equivalent to the proof of (61) on page 93. As mentioned there, this method can be regarded as a true measurement of decrement, since the current flowing in the circuit has the natural decrement of the circuit. It is difficult to obtain high accuracy by the method in practice because of the difficulty of obtaining pure impulse excitation. An example of such a measurement is given on page 190 below.

The method is specially convenient when an instrument is used in which the deflection d is proportional to the current squared. Then (84) becomes

$$R = R_1 \frac{d_1}{d - d_1} \tag{85}$$

This is still further simplified if the resistance R_1 is adjustable, so that d_1 can be made equal to one-half d. The equation then reduces to

$$R = R_1 \tag{86}$$

This is commonly known as the half-deflection method.

Use of Damped Excitation.—The resistance variation method has already been shown to be usable with either undamped or free oscillations. It can also be used when the source of emf is damped so that both forced and free oscillations exist in the circuit. Equation (62) on page 94 above gives the relation between the currents and the decrements of the circuits. As there stated, such a measurement may be looked upon as a direct measurement of decrement. It is possible to calculate either the decrement δ' of the supplied emf as in equations (63) to (65) when the constants of the measurement circuit are known, or the decrement δ or resistance R of the measuring circuit when δ' is known.

A convenient form of the solution for R is obtained from equation (66) and the relations $\delta = \tau \dfrac{R}{L\omega}$ and $\delta_1 = \tau \dfrac{R_1}{L\omega}$,

$$R = R_1 \frac{KI_1^2}{I^2 - KI_1^2} \qquad (87)$$

where

$$K = 1 + \frac{\delta_1}{\delta' + \delta} \qquad (88)$$

This is, of course, not an explicit solution for R, since K involves δ and therefore R, but gives a ready means of finding R or δ when the sum of the two decrements $\delta' + \delta$ is known from some other measurement, such as the reactance-variation method described below. Thus, a combination of the two methods gives both δ' and δ, or δ' and R.

There are two interesting special cases in which the measurement is simplified. When the decrement δ' of the supplied emf is very small and is negligible compared with δ, equation (87) reduces to

$$R = R_1 \frac{I_1}{I - I_1} \qquad (89)$$

identical with (80) above, the equation for the use of undamped emf. This is to be expected, since undamped emf is the limiting case of small decrement. When, on the other hand, δ and δ_1 are both very small compared with δ', K becomes unity and equation (87) reduces to

$$R = R_1 \frac{I_1^2}{I^2 - I_1^2} \qquad (90)$$

This happens to be the same as equation (84) above, the equation for the use of impulse excitation. The proof given here can not, however, be regarded as a deduction of equation (84) for impulse excitation, as it has been by some writers, since Bjerknes' equation (p. 187) is involved, which assumes that δ' and δ are both small.

50. REACTANCE VARIATION METHOD

This has been called the decrement method, a name which is no more applicable to this than to the other methods of resistance measurement since all measure decrement in the same sense that this does. That the method primarily measures resistance rather than decrement is seen from the fact that in its simple and most accurate form it utilizes undamped current, which has no decrement.

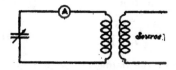

Fig. 128.—*Circuit for measuring resistance by the reactance variation method*

The method is analogous to the resistance-variation method, two observations being taken. The current I_r in the ammeter is measured at resonance, the reactance is then varied and the new current I_1 is observed. The total resistance of the circuit R is calculated from these two observations. The reactance may be varied by changing either the capacity, the inductance, or the frequency, the emf being maintained constant. The reactance is zero at resonance and it is changed to some value X_1 for the other observation. With undamped emf E, the currents are given by

$$I_r^2 = \frac{E^2}{R^2}$$

$$I_1^2 = \frac{E^2}{R^2 + X_1^2}$$

From these it follows that

$$R = X_1 \sqrt{\frac{I_1^2}{I_r^2 - I_1^2}} \qquad (91)$$

This has a similarity to $R = R_1 \dfrac{I_1}{I - I_1}$, the equation (80) for the resistance-variation method. It is also interesting that when the reactance is varied by such an amount as to make the quantity under the radical sign equal to unity, the equation reduces to

$$R = X_1 \tag{92}$$

This is similar to $R = R_1$, which is the equation for the quarter-deflection and half-deflection resistance variation methods.

Special Cases of Method.—When the reactance is varied by changing the setting of a variable condenser, the equation (91) becomes (27) given on page 38. For variation of the inductance, (91) becomes

$$R = \pm \omega (L - L_r) \sqrt{\frac{I_1^2}{I_r^2 - I_1^2}} \tag{93}$$

For variation of the frequency, (91) becomes

$$R = \frac{\pm L (\omega^2 - \omega_r^2)}{\omega} \sqrt{\frac{I_1^2}{I_r^2 - I_1^2}} \tag{94}$$

It must be noted that variation of the frequency requires some alteration in the source of emf, and the greatest care is necessary to insure that the condition of constant emf is fulfilled.

A convenient method which differs slightly from those just described is to observe two values of the reactance both corresponding to the same current I_1 on the two sides of the resonant value I_r. For observation in this manner of two capacity values C_1 and C_2,

$$R = \frac{1}{2\omega} \frac{C_2 - C_1}{C_2 C_1} \sqrt{\frac{I_1^2}{I_r^2 - I_1^2}} \tag{95}$$

Decrement Calculation.—It is often convenient to calculate directly the decrement of the circuit instead of the resistance. Formulas for decrement exactly corresponding to the resistance formulas already given are obtained by application of the simple relations between resistance and decrement and are the same as formulas (96) to (100) below with δ' omitted. The formulas thus obtained are rigorous, as are the foregoing resistance formulas, for undamped emf, and hold with sufficient accuracy for damped emf when the damping is negligibly small.

When the damping of the supplied emf is appreciable, the same procedure is followed in making the measurement, and the equations are only slightly different. When the emf is supplied by coupling to a primary circuit in which current is flowing with a decrement δ', Bjerknes' classical proof shows that the sum of the primary and secondary decrements is given by the same expression as that which gives the decrement δ of the secondary when the emf is undamped. Thus (27), (93), (94), and (95) correspond to

$$\delta' + \delta = \pi \frac{\pm (C_r - C)}{C} \sqrt{\frac{I_1^2}{I_r^2 - I_1^2}} \tag{96}$$

$$\delta' + \delta = \pi \frac{\pm (L - L_r)}{L_r} \sqrt{\frac{I_1^2}{I_r^2 - I_1^2}} \tag{97}$$

$$\delta' + \delta = \pi \frac{\pm (\omega^2 - \omega_r^2)}{\omega \omega_r} \sqrt{\frac{I_1^2}{I_r^2 - I_1^2}} \tag{98}$$

$$\delta' + \delta = \pi \frac{C_2 - C_1}{C_2 + C_1} \sqrt{\frac{I_1^2}{I_r^2 - I_1^2}} \tag{99}$$

Formula (98) is also equivalent to

$$\delta' + \delta = \pi \frac{\pm (\lambda_r^2 - \lambda^2)}{\lambda_r \lambda} \sqrt{\frac{I_1^2}{I_r^2 - I_1}} \tag{100}$$

These formulas are correct only when: (1) The coupling between the two circuits is so loose that the secondary does not appreciably affect the primary, (2) δ' and δ are both small compared with 2π, and (3) the ratio $\frac{(C_r - C)}{C}$ and the corresponding ratios are small compared with unity.

In any of these methods the calculation is obviously simplified if the reactance is varied by such an amount as to make $I_1^2 = \frac{1}{2} I_r^2$. This is done very easily when the current measuring instrument is graduated in terms of current squared. The quantity under the square-root sign in all the preceding equations becomes unity, greatly simplifying the formulas. A still further simplification by which all calculation is eliminated is utilized in special decremeters as described below, section 55.

51. RESISTANCE OF A WAVE METER

The accurate measurement of resistance or decrement of a wave-meter circuit is of first importance because the wave meter is frequently used to measure other resistances and the decrements of waves. It is the calibration of a resistance-measuring standard. Several forms of the resistance-variation and the reactance-variation methods may be used.

The resistance of a wave meter is, of course, not a single, constant value. It varies with frequency and with the detecting or other apparatus connected to the wave-meter circuit. Usually both the resistance and the decrement of the circuit vary with the condenser setting. It is usually desirable to express either resistance or decrement in the form of curves for the several wave-meter coils, each for a particular detecting apparatus or other condition. An example of such a curve is given on page 190.

Resistance Variation.—Any of the forms of the resistance-variation method may be used. The apparatus and procedure are the same in all cases. The wave-meter coil is loosely coupled to the source. The current is read on the indicating device shown schematically as A in Fig. 129. A resistance standard of the type already described is then inserted at R_1 and the current read

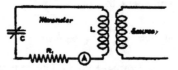

Fig. 129.—*Measurement of wave meter resistance*

again. The calculation of resistance depends on the damping of the source and the kind of current-measuring device.

When a pliotron, arc, or other source of undamped waves is used, formula (80) above is used. When the current-measuring device is a current-square meter, thermocouple or crystal detector with galvanometer, or other apparatus which is so calibrated that deflections are accurately proportional to the square of the current, and when in addition a continuously variable resistance standard is used, the quarter-deflection method may be employed eliminating all calculation. As explained on page 182, the resistance of the circuit is equal to the inserted resistance required to reduce the deflection to one-quarter.

When a buzzer or other damped source is used, some auxiliary measurement or special condition is needed, in order to evaluate or eliminate the decrement of the source. If this decrement is known, the decrement or resistance of the wave-meter circuit may be obtained. The solution is, however, complicated and, as a matter of fact, this method in not used. Instead of δ', the decrement of the source, being known explicitly, the more usual case is that $\delta' + \delta$, the sum of the decrements of source and wave meter, is known from a measurement by the reactance-variation method. The wave-meter resistance is then calculated by

$$R = R_1 \frac{K I_1^2}{I^2 - K I_r^2}$$

where $K = 1 + \frac{\delta_1}{\delta' + \delta}$, and resistance and decrement are related by $\delta = \frac{\pi R}{\omega L}$ and the similar formulas given on page 316. The calculation is considerably simplified in the two special cases of δ' very small or δ very small; formulas (98) and (99), respectively, apply. The latter is identical with the equation for impulse excitation, but with that exception these methods may be used only when δ' and δ are both small.

For impulse excitation from a buzzer or other source, equation (84) is used to calculate the resistance. When the current indicator is calibrated in terms of the square of current and the resistance standard is continuously variable, the measurement is conveniently made by the half-deflection method. In this case the resistance of the circuit equals the inserted resistance.

Reactance Variation.—This method may be used with either a damped or undamped source. When the emf is undamped or of extremely small damping, formulas (27) and (93) to (95) apply. It is customary to reduce the labor of computation by varying the reactance by such an amount that $\sqrt{\frac{I_1^2}{I_r^2 - I_1^2}} = 1$, in which case

$$R = \frac{\pm (C_r - C)}{\omega C_r C} \tag{101}$$

$$R = \pm \omega (L - L_r) \tag{102}$$

$$R = \frac{\pm L (\omega^2 - \omega_r^2)}{\omega} \tag{103}$$

$$R = \frac{1}{2\omega} \frac{C_2 - C_1}{C_2 C_1} \tag{104}$$

When damped emf is used, formulas (96) to (100) apply. They also are simplified in practice by making $\sqrt{\frac{I_1^2}{I_r^2 - I_1^2}} = 1$.

They require either that δ', the decrement of the source, be known, or that another relation be obtained between δ' and δ by an independent measurement. It is not often that a source of fixed, known decrement is maintained in a laboratory, as the decrement varies with frequency and every other condition of use. Hence, the usual procedure is the combination of this measurement with a resistance-variation measurement as described above.

An example of measurement of wave-meter resistance expressed in terms of decrement is given in Fig. 130. This shows the results of two independent measurements, one by the resistance-variation

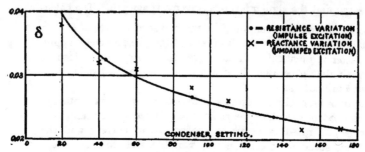

Fig. 130.—*Variation of the decrement of a wave meter with condenser setting*

method, using impulse excitation and equation (84), and another by the reactance-variation method employing equation (27).

52. RESISTANCE OF A CONDENSER

The methods ordinarily used for measurement of resistance of a condenser or of an inductance coil require a variable condenser whose effective resistance must be either negligibly small or accurately known. This condenser is used to retune the circuit to resonance after the unknown is taken out of the circuit. The standard condensers of negligible resistance used at the Bureau of Standards are described on page 119. These measurements may be made with an ordinary wave meter, provided the resistance of the circuit is accurately known for different condenser settings.

Simple Methods.—The simplest method is that of substitution. The condenser to be tested is connected in series with a coil and an ammeter of some sort, and loosely coupled to a source. The condenser is then replaced by the standard condenser and a series resistance. The resistance required to make the deflection at resonance the same as before is taken as the resistance of the condenser. This method is not very accurate, because the change of condensers changes the emf's electrostatically induced in the circuit.

Another method utilizes the principle of reactance variation. The frequency supplied to the condenser circuit is varied by changing the setting of a condenser in the supply circuit. Under certain conditions, equation (27) or (95) applies. The method can be made to give phase differences directly by use of a suitable scale of phase differences on the condenser in the supply circuit. (See sec. 55.)

Precision Method.—Accurate measurements may be made by the resistance-variation method. The circuit is tuned to resonance by varying the frequency supplied, and the total resistance of the circuit is measured, with the unknown condenser in circuit. It is replaced by the standard condenser, the setting of which is varied until resonance is obtained, and then the resistance of the circuit is measured again. The difference of the two measured values is

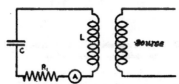

Fig. 131.—*Circuit used for measurement of high-frequency resistance of a condenser*

the resistance of the condenser under test. As previously noted, precautions are necessary to avoid changing the stray electrostatic emf's in the circuit when the resistance R_1 is introduced. In respect to this it has been found desirable to insert R_1 between the condenser and the current-measuring device, and, if a ground wire is used, to connect it to the shielded side of the condenser or to the ammeter case. Also, the coupling must be loose enough so that too much power is not withdrawn from the source.

The manipulation is made more convenient by using a double-throw switch to place the two condensers in circuit. The base of the switch must be a material which has very small phase difference; paraffin has been found suitable. The resistance standards, when in the shape of short links, may be used as part of this switch, as shown at R_1 in Fig. 132. Another refinement of the measurement is to place a small variable condenser C_1 of negligible resistance in parallel with the inductance coil. This gives a fine adjustment to resonance.

Example.—An example of a measurement at one frequency with a pliotron as a source of undamped emf is given below. The condenser is a fixed condenser with molded dielectric. The column headed "d" gives deflections of galvanometer attached to

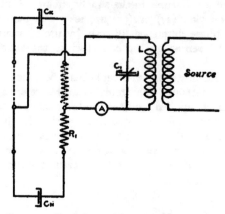

FIG. 132.—*Circuit for precision measurement of condenser resistance with switching device and small tuning condenser*

a thermocouple; deflections are proportional to the square of the current.

TABLE 4.—Observations on Resistance of a Mica Condenser

[C=0.00406 µl, L=40 µh, λ=760 m, R_X=R_N=0.09 ohm=resistance of condenser X; phase difference= $6.5\dfrac{0.09(4.060)}{760}$=3'.]

C	R_1	Galvanometer				$\sqrt{\frac{d}{d_1}-1}$	$R=\dfrac{R_1}{\sqrt{\frac{d}{d_1}-1}}$
		Zero	Deflection	d	Mean d		
	0	13.95	45.70	31.75	31.89	....	
	0.503	13.95	31.68	17.73	17.78	0.340	1.480
X......	.810	13.92	27.30	13.38	13.30	.548	1.479 }1.47r
	1.042	13.90	24.82	10.92	10.92	.708	1.472
	.810	13.95	27.18	13.23	....	....	
	.503	13.90	31.75	17.83	....	....	
	0	13.92	45.95	32.03	....	....	
	0	13.98	48.58	34.60	34.58	....	
	.503	13.92	32.45	18.53	18.62	.362	1.390
	.810	13.92	27.72	13.80	13.81	.582	1.391 }1.38o
N.....	1.042	13.88	25.15	11.27	11.27	.752	1.387
	.810	13.88	27.70	13.82	....	....	
	.503	13.90	32.60	18.70	....	....	
	0	13.95	48.50	34.55	....	....	

The resistance of a condenser is generally measured at several frequencies. If the resistance is mainly due to dielectric absorption, the resistance is generally inversely proportional to frequency. Variable condensers are usually measured at several settings. For a variable air condenser with semicircular plates having a small resistance mainly due to dielectric absorption in the separating insulators, the resistance is inversely proportional to the square of the setting.

53. RESISTANCE OF A COIL

The resistance of a coil depends upon its position in the circuit, i. e., whether the emf acting upon the circuit is impressed in the coil itself or at some other point in the circuit. This difference arises from the effects due to the capacity of the coil. When the emf is induced in the coil itself, the capacity of the coil is to be considered in series with the inductance of the coil but in parallel

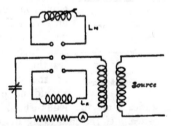

Fig. 133.—*Circuit for measuring high-frequency resistance of a coil by the substitution method*

with the rest of the circuit. When, however, the emf is impressed at some other point in the circuit, the coil capacity and inductance are in parallel with each other. When the resistance of a coil is measured by any of the methods given in this section, the value of the resistance obtained is valid for the coil only when used in the same position relative to the driving emf.

The simplest method is that of substitution. In a wave meter circuit coupled to a source of emf by an independent coupling coil, the deflection is first observed with the coil whose resistance is desired in circuit. If the wave meter condenser resistance is known for various settings, the coil is then replaced by a standard coil whose resistance is known, the condenser retuned to resonance, and resistance inserted until the deflection is the same as before. The change of coils may be made by a double-throw switch. If a variable inductor of known resistance is available, the procedure is still simpler, as the condenser setting need not be changed; the coil under test is replaced by the variable inductor, which is used to obtain resonance, and then resistance is inserted to equalize deflections. As in previous cases, the substitution method is not an accurate one, but is valuable for speedy determination of approximate values.

More accurate measurements may be made by either the resistance-variation or the reactance-variation method. Either of these methods may be used to determine the resistance in each of the three following procedures.

Known Circuit.—The circuit consists of the unknown coil, a condenser, the current-indicating instrument, and connecting leads. The emf is introduced into the circuit by coupling to the unknown coil. The resistance of the total

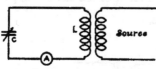

FIG. 134.—*Coil resistance measurement in terms of known circuit or standard coil*

circuit is obtained by the resistance-variation or the reactance-variation method and the coil resistance determined by subtracting the resistance of the rest of the circuit. The condenser should be practically perfect or of known resistance, and the leads should be of negligible or calculable resistance. If the indicating instrument is a thermo-element or current-square meter with fine wire heating element, its resistance may be determined with direct current. In doubtful cases it can be determined at the frequency of the measurement by a separate experiment as outlined on page 179. Thermo-elements of low resistance, as described on page 157, are especially suited for this measurement, since the precision is higher if the resistance of the indicating device is a small part of the total resistance of the circuit.

Known Coil.—Two resistance measurements at high frequency are required when the unknown coil is compared with a standard coil of known resistance. The inductance of the standard coil should be of the same order of magnitude as that of the unknown, but need not be equal to it.

The standard is substituted for the unknown and the condenser varied for resonance and the resistance of the circuit obtained in each case.

Auxiliary Coil.—The third procedure is to measure the resistance of a simple radio circuit consisting of a condenser, ammeter, and any coil, all in series, and then insert the coil to be tested, retune to resonance, and measure the resistance again. The coil must be inserted at such a place in the circuit that it has no mutual inductance with the auxiliary coil, and so that there is no emf induced in it by the source.

54. DECREMENT OF A WAVE

Any measurement of the resistance of a circuit by the methods already given is in a sense a measurement of decrement, since it enables calculations of the decrement which the circuit would have when oscillating freely. In particular, the use of impulse excitation with the resistance-variation method measures the decrement of the oscillations actually existing in the measuring circuit during the measurement, and therefore the decrement of the wave emitted by the measuring circuit.

There is a class of decrement measurements entirely apart from the measurement of decrement or resistance of a circuit. This is the determination of decrement of an emf, due either to a nearby antenna or other circuit or to a wave traveling through space. The fundamental principles of decrement measurement have been given in section 26 above. A simple wave-meter circuit is placed so as to receive the wave and a decrement measurement is made by either the resistance-variation or the reactance-variation method. If the resistance of the wave-meter circuit is known for the frequency and other conditions of the measurement, the decrement of the wave is calculated. For the resistance-variation method δ', the decrement of the wave, is given by formula (63), page 63, or one of the simplified formulas (64) or (65).

For the reactance-variation method the decrement is given by (96) to (100), page 187, or, in case the simplified procedure is followed, making $I_1^2 = \frac{1}{2}I_r^2$.

$$\delta' + \delta = \pi\frac{\pm(C_r - C)}{C} \tag{105}$$

$$\delta' + \delta = \pi\frac{\pm(L - L_r)}{L_r} \tag{106}$$

$$\delta' + \delta = \pi\frac{\pm(\omega^2 - \omega_r^2)}{\omega\,\omega_r} \tag{107}$$

$$\delta' + \delta = \pi\frac{C_2 - C_1}{C_2 + C_1} \tag{108}$$

$$\delta' + \delta = \pi\frac{\pm(\lambda_r^2 - \lambda^2)}{\lambda_r\lambda} \tag{109}$$

The measurement may also be made by a direct-reading decremeter (see next section), and $(\delta' + \delta)$ read directly from the scale of the instrument.

When the resistance or decrement of the wave-meter circuit is not known, measurements are made by both methods: and the combination of the two yields the value of δ', making use of equation (63).

55. THE DECREMETER

A decremeter is a wave-meter conveniently arranged for measurements of resistance or decrement. The forms usually employed make use of the reactance-variation method. Ways have been devised for manipulating the instrument in such a way that the decrement may be read directly from a scale. While, of course, resistance can be calculated from a measured value of decrement, the principle application of the decremeter is in the measurement of the decrement of a wave.

Another important use is in the measurement of phase difference of a condenser. Since the decrement due to a condenser is π times its phase difference, a measurement of decrement gives directly the phase difference; if desired, the scale may be calibrated in terms of phase difference instead of decrement.

Determination of the Scale of a Decremeter.—Any wave meter whose circuit includes some form of ammeter may be fitted with a special scale from which decrements may be read directly. The procedure for a wave meter having any sort of variable condenser is given here.

In the usual use of the reactance variation method of determining decrement, the current I_r is observed when the condenser is adjusted to the value C_r to produce resonance, and the condenser is then changed to another value C and the current I_1 read. When the second condenser setting is such that the $I_1^2 = \frac{1}{2} I_r^2$, the decrement is calculated by

$$\delta' + \delta = \pi \frac{\pm (C_r - C)}{C}$$

A certain value of decrement therefore corresponds to that displacement of the condenser's moving plates which varies the capacity by the amount $(C_r - C)$. The displacement for a given decrement will in general be different for different values of C, the total capacity in the circuit. At each point of the condenser scale, therefore, any displacement of the moving plates which changes the square of current from I_r^2 to $\frac{1}{2} I_r^2$ means a certain value of $(\delta' + \delta)$.

A special scale may therefore be attached to any condenser with graduations upon it and so marked that the difference between the two settings, when the square of current is I_r^2 and $\frac{1}{2} I_r^2$, is equal to the decrement. The spacing of the graduations at different parts of the scale depends upon the relation beween capacity and displacement of the moving plates. When this relation is known, the decrement scale can be predetermined. A scale may therefore be fitted to any condenser, from which decrement may be read directly, provided the capacity of the circuit is known for all settings of the condenser. The decrement scale may be attached either to the moving-plate system or to the fixed condenser top. It is usually convenient to attach it to the unused half of the dial opposite the capacity scale. The value of decrement determined by this method is $(\delta' + \delta)$, where δ is the decrement of the instrument itself. This must be known from the calibration of the instrument (e. g., as in sec. 51), the value of δ', the decrement of the wave under measurement, being then obtained by subtraction.

Simple Direct-Reading Decremeter.—It is particularly easy to make a decremeter out of a circuit having a condenser with semi-circular plates. Such condensers follow closely the linear law,

$$C = a\theta + C_0,$$

where θ is the angle of rotation of the moving plates and a and C_0 are constants. It can be shown that the decrement scale applicable to such a condenser is one in which the graduations vary as the logarithm of the angle of rotation. Furthermore, the same decrement scale applies to all condensers of this type. This scale has been calculated and is given in Fig. 135. It is calculated to fit equation (108), for observations on both sides of resonance, and not for equation (105). This scale may be used on any condenser with semicircular plates. The scale may be cut out and trimmed at such a radius as to fit the dial and then affixed to the condenser with its O point in coincidence with the graduation which corresponds to maximum capacity. This usually puts it on the unused half of the dial opposite the capacity scale. If the figures are trimmed off, they can be added over the lines in red ink. If it is desired not to mutilate the page, the scale may be copied. This scale gives accurate results if the capacity scale is so set that its indications are proportional to the capacity in the circuit.

A measurement of decrement is made by first observing the current-square at resonance, then reading the decrement scale at a setting on each side of resonance for which the current-square

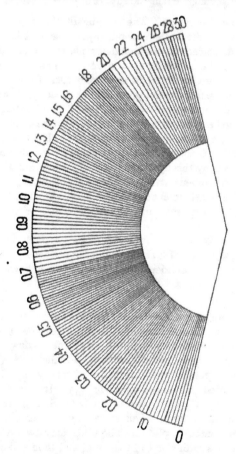

Fig. 135.—*Direct-reading decrement scale for any semicircular-plate condenser*

is one-half its value at resonance. The difference between the two readings on the decrement scale is the value of $\delta' + \delta$.

The scale permits accurate measurement of fairly large decrements, but offers no precision in the measurement of very small decrements, particularly at the low-capacity end of the scale.

A similar scale is readily made to read phase differences directly. The readings of Fig. 135 are all multiplied by 18.24, to give phase differences in degrees. The instrument is then specially valuable in measuring the phase differences of condensers.

Decremeter with Uniform Decrement Scale.—Just as it is possible to determine a decrement scale to fit a condenser having any sort of law of capacity variation, it is equally possible to design a condenser with capacity varying in such a way as to fit any specified decrement scale. A uniform decrement scale—i. e., one in which the graduations are equally spaced—is particularly convenient, and is the kind used in the Kolster decremeter. A uniform decrement scale requires that the condenser plates be so shaped that for any small variation of setting the ratio of the change in the capacity to the total capacity is constant. The condenser required to give this uniform scale has its moving plates so shaped that the logarithm of the capacity is proportional to the angle of rotation of the plates. This condenser is discussed above on page 116.

The decremeter is fully described in Bulletin of the Bureau of Standards, 11, page 421; 1914 (Scientific Paper No. 235). A view of the inside arrangement is shown in Fig. 218, facing page 320, of this circular. The spiral shape of the condenser moving plates is shown. Figures 219 and 220, facing page 32, show different forms in which the instrument is made. The decrement scale is not attached directly to the moving plates, but is on a separate shaft geared to the moving plates at a 6 to 1 ratio. The decrement scale is thus opened out, so that very precise measurenemts may be made.

To measure decrement the condenser setting is first varied to obtain maximum deflection of the current-square meter, and then varied until half this deflection is obtained. The movable decrement scale is then set at zero and clamped to its shaft, and the condenser setting is varied until the same deflection is obtained on the other side of the maximum. The reading on the decrement scale is the value of decrement sought.

This decremeter is used in the inspection service of the Bureau of Navigation, Department of Commerce, and by radio engineers in the Army and Navy and elsewhere.

SOURCES OF HIGH-FREQUENCY CURRENT

56. ELECTRON TUBES

For the purposes of measurement as well as in the transmission of radiograms, sources which furnish undamped currents are coming into more general use than those which supply damped oscillations. Thus the various forms of spark, used almost exclusively in the past, are giving way to the electron-tube generator, the arc, and the high-frequency alternator. These sources are used both in radio communication and in laboratory work, but for the latter the electron tube is preeminently the best. When the source furnishes undamped current many methods of measurement are simplified, and since the sharpness of tuning is increased, a higher precision is obtainable in methods which depend upon tuning to resonance.

Since most of the measurements at radio frequencies are based upon deflection methods, the primary requirement of a source for such measurements is that the intensity and frequency of the generated current be constant. It is only within the past few years that an almost ideal source has become available, viz, electron-tube generators, such as the pliotron, audion, oscillion, etc. These electron tubes consist of an evacuated bulb containing the following three elements:

(a) A heated filiament which acts as a source of electrons.

(b) A metal ''plate'' placed near the electron source. (Across the plate and filament, outside of the bulb, is connected a battery, so that an electron current flows from filament to plate.)

(c) A "grid" consisting of fine wire or of a perforated metal sheet placed between plate and filament so that the electrons have to pass through the grid to get from filament to plate.

Using these sources, undamped high-frequency current can be obtained which is as steady as the current from a storage battery and strictly constant in frequency. Furthermore, these generators are extremely flexible as to the frequencies which can be obtained, the same tubes having been used to generate currents ranging in frequency from 1 cycle in two seconds to 50 000 000 cycles per second.[25] A high-frequency output of 500 watts or more has been attained with a single tube. For ordinary measuring purposes with sensitive indicating devices, such as a thermo-element and galvanometer, about 5 watts of high-frequency

---

[25] See reference No. 138, Appendix 2.

output are sufficient. When it is desired to use low-range hot-wire ammeters, 10 to 20 watts may be required.

On account of the extreme importance of the three-electrode tubes, both as generators and as detectors, and since the full realization of their utility and a satisfactory explanation of their functioning are of recent date, it is worth while to outline rather fully the phenomena upon which their operation is explained.

Thermionic Emission.—The modern conception of current flow in metals assumes that the conduction of electricity consists in the motion of electrons (see p. 8) under the action of an applied electromotive force. When not acted upon by an external emf these small negatively charged particles move about in the metal in zigzag paths in all directions, colliding with the atoms of the metal. Their mean velocity of motion depends upon the temperature, increasing with the temperature. At the surface of the metal, according to the theory of Richardson,[20] the electrons are restrained from leaving the metal by electric forces entirely similar to the molecular forces which cause the surface tension of a liquid. Further, just as in the evaporation of a liquid, a certain number of electrons will in each second attain a high enough velocity to escape from the metal, and since the mean velocity increases with the temperature the number of electrons escaping per second will increase with the temperature. The heated filament in the audion or pliotron is in this manner a source of electrons. The withdrawal of the negative electrons from the heated filament leaves it positively charged, thus tending to draw them back again and a state of equilibrium may be attained in which the same number are drawn back per second as are being emitted. When, however, a body maintained at a positive potential relative to the filament is brought into the field a certain proportion of the electrons will be attracted to the positive body and constitute a current of electricity between the filament and the positively charged body. In the electron tube this body is the "plate" and its potential is maintained positive with respect to the filament by a battery commonly called the B battery. If in the case of a tube with an extremely high vacuum the voltage of the B battery is increased, the flow of electrons or "plate current" will increase up to a point where practically all of the electrons emitted by the filament are being drawn over to the plate. If, on the other hand, the plate voltage is kept constant and the filament temperature increased, thus increasing the

[20] See reference No. 131, Appendix 2.

number of electrons emitted per second, the plate current will also increase up to a certain temperature, but beyond this temperature will remain practically constant even though more electrons are being given off. The explanation of this behavior [17] is that the stream of negative electrons flowing through the tube acts as a space charge of negative electricity which neutralizes the field due to the positive plate. In consequence only a limited number of electrons can flow to the plate per second with a given plate voltage, and the remainder are compelled to return to the filament again.

Grid Control.—If in any way this space charge is neutralized, there will be an increase in the plate current; on the other hand,

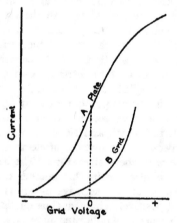

FIG. 136.—*Variation of plate current (usually in milliamperes) and grid current (usually in microamperes) with grid voltage*

anything that will aid the space charge will result in a decrease in the plate current. In the audion or pliotron these effects are brought about by the grid of wires between the plate and filament. If this grid is charged positively with respect to the filament, the effect of the space charge will be neutralized to an extent depending upon the charge on the grid, and the electron current through the tube will increase until the field due to the grid charge is also neutralized by the space charge.[28] Some few electrons will strike the

---

[17] See reference No. 133, Appendix 2.

[28] In tubes which are not evacuated to a high degree, the residual gas may become ionized and markedly affect the behavior of the tube. The ionization of the gas tends to neutralize the space charge, thus permitting larger currents to pass through the tube. To a certain extent such ionization is of value in the use of the tube as a detector, though when the ionization becomes intense and the tube shows a blue glow, so large a current passes through the tube that it is unaffected by variations of the grid voltage and its detecting qualities are lost.

grid wires and there will result a flow of current in the grid circuit, but this will, in general, be small relative to the plate current. If the grid is charged negatively with respect to the filament, the charge on the grid will then aid the space charge in driving the electrons back to the filament, resulting in a lowering of the plate current. In this latter case the number of electrons striking the grid will be very small and in consequence practically no current will flow in the grid circuit. The control of the plate current by the grid voltage and also the dependence of the current in the grid circuit upon the grid voltage are shown in curves of Fig. 136. Curve *A* shows the current in the plate circuit when the *B* battery is kept constant, but different voltages are applied between the grid and that terminal of the filament to which the negative of the filament battery is connected. Curve *B* represents on a magnified

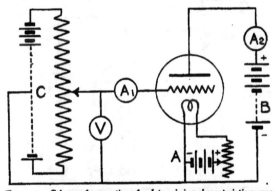

Fig. 137.—*Scheme of connections for determining characteristic curves*

scale the current in the grid circuit for different voltages of the grid with respect to the negative terminal of the filament. The ordinates and shape of these so-called characteristic curves depend upon a number of factors, such as *B* battery voltage, fineness and spacing of the grid wires, location of the grid relative to the other elements, etc. Fig. 137 shows the scheme of connections which may be used in determining such curves. The ammeter A_1 measures the current in the grid-filament circuit and A_2 measures the plate current. By means of the sliding contact on the shunt resistance to the battery *C*, the voltage between the filament and grid may be varied and made positive or negative, the voltage being read by the voltmeter V. The *B* battery voltage is held constant while the curve is taken.

57. ELECTRON TUBE AS DETECTOR AND AMPLIFIER

As Detector of Damped Oscillations.—A single tube may perform separately or simultaneously the functions of a detector, amplifier, and generator. It will first be considered as a simple detector of damped oscillations. The circuits shown in Fig. 138 indicate one possible way of using the tube as a detector. The circuit LC_1 is tuned to the oscillations in the antenna A. The C battery with variable resistance permits the adjustment of the grid potential with respect to the filament, so that the tube may be worked at any point on the characteristic curve of the plate current. Suppose that this voltage is adjusted to correspond to the

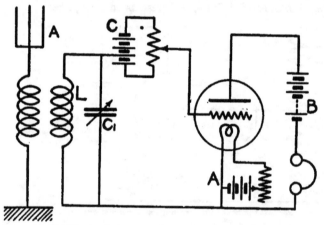

FIG. 138.—*Possible circuits for using the electron tube as a detector of damped oscillations*

point X-(Fig. 139) where the change in slope of the curve is large. If now a train of oscillations is set up in the antenna and hence in the secondary circuit, the alternating voltage across the condenser terminals will be superimposed upon the steady voltage of the C battery. It will be seen from the characteristic curve that an increase in voltage, say, from a to b, produces a large increase in the plate current (i. e., from x to y), while a decrease in voltage of the same amount from a to c produces a much smaller change (from x to z) in the current. Thus, as the result of a wave train such as (1) in Fig. 140, the plate current will be changed about its normal value in some such way as (2) which is equivalent to a resultant increase in plate current. This increase of plate current during a train of waves gives rise to a pulse of current in the telephone as shown in (3). This pulse will act upon the telephone diaphragm, and if the wave trains and hence the pulses in the telephone current are arriving at the rate of 1000 per second (corresponding to the spark frequency at the transmitting station), a 1000-cycle note will be heard in the phone. This use of the tube as a detector is entirely similar to the use of a crystal detector.

Condenser in Grid Lead.—If a condenser is inserted instead of the C battery in the lead to the grid, as C_2 in Fig. 141, the behavior

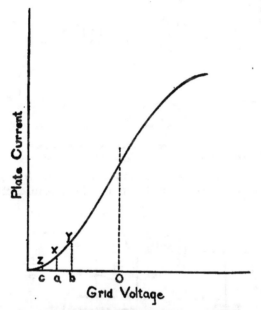

Fig. 139.—*Plate characteristic, showing region of curve where rectifying action is large*

of the tube as a detector of damped [20] oscillations is altered and depends to a great extent upon the characteristic curve of the grid current. The grid is insulated from the filament by the condenser C_2, excepting for such leakage as may take place through this condenser or in or about the tube. Suppose first that the tube is put into operation with the grid and filament at the same potential and with no incoming oscillations. It will be seen from

[20] Although damped oscillations are referred to here and in the usual treatments of the subject, the same considerations apply to undamped oscillations which are periodically interrupted either in the transmitting or receiving circuits so that the tube receives groups or trains of waves.

the filament-grid curve that there will be a flow of electrons to the
grid and the grid will become negative with respect to the filament,

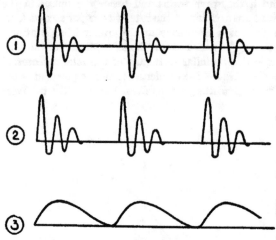

Fig. 140.—*Action of the electron tube as a detector:* (1) *Incoming oscillations,* (2) *variations in plate current,* (3) *effective telephone pulses*

thereby reducing the flow to itself and to the plate until the leakage
away from the grid is equal to the flow to it. In some tubes the

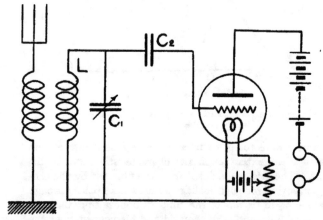

Fig. 141.—*The electron tube as a detector of damped oscillations, using a condenser in the grid circuit*

grid may be so highly insulated that it accumulates a negative
charge sufficiently high to reduce the plate current practically to

zero. In such cases it is necessary to provide an artificial leak through a high resistance across C_2.

Suppose now that the grid has attained its equilibrium potential and the plate current its corresponding value and a series of wave trains impinges upon the antenna as in (1) of Fig. 142. The oscillations in the circuit LC_1 will cause the grid potential to oscillate about its normal value. When the grid becomes positive

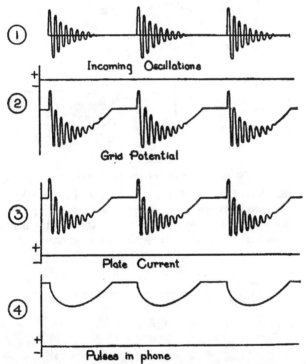

FIG. 142.—*Action of the electron tube as a detector connected as in Fig. 141*

relative to its normal value there will be a considerable increase in the flow of electrons to it, overbalancing the reduction in the flow when on the negative half of the wave. Thus, during a wave train the grid will accumulate a negative charge and its mean potential will be lowered, as in (2) of Fig. 142. In consequence the mean plate current will be reduced. However, between wave trains the excess charge on the grid will leak off, restoring the plate current to its normal value. This is shown in (3) of Fig. 142. Each wave train will produce a reduction in the current through the phones as in (4) of the same figure and a note corresponding to the wave train frequency will be heard.

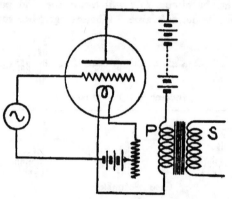

FIG. 143.—*Use of electron tube as an amplifier*

Amplification.—If, as in Fig. 143, a source of alternating emf were interposed between the filament and grid of an audion or pliotron, the potential of the grid with respect to the filament would alternate in accordance with the alternations of the

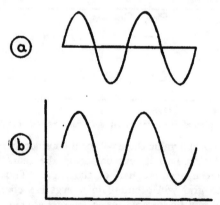

FIG. 144.—*Variations of plate current with grid voltage*

generator. These variations of the grid potential produce changes in the plate current corresponding to the plate characteristic. If the mean potential of the grid and the amplitude of its alternations are such that the plate current is always in that portion of

its characteristic where it is a straight line, then the alternations
of the grid potential will be exactly duplicated in the variations
of the plate current and the latter will be in phase with the former,
at least in a high vacuum tube. ¯ Thus, if (a) of Fig. 144 represents
the alternating potential of the grid, then (b) would represent the
fluctuations of the plate current. For a given amplitude in (a),
the amplitude of the alternating component in (b) will depend
upon the steepness of the plate characteristic, increasing with
increasing slope. The alternator in the grid lead supplies only the
very small grid-filament current, thus the power drawn from it is
extremely small. The power represented by the alternating
component of the plate current is, however, considerable; thus
there is a very large power amplification. This larger source of

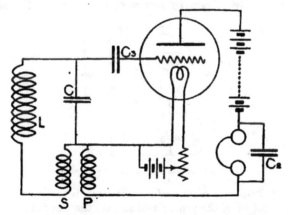

Fig. 145.—*Use of electron tube as a regenerative amplifier*

power might be utilized by inserting the primary *P* of a transformer in the plate circuit, as in Fig. 143, in which case the alternating component alone would be present in the secondary *S*.
This illustrates the principle of a vacuum tube as a relay. The
voltage in *S* might again be inserted in the grid lead of a second
vacuum tube and with proper design a further amplification obtained in the plate circuit of the second tube. This may be
carried through further stages and illustrates the principle of
multiple amplification.

Regenerative Amplification.—It has been shown by E. H.
Armstrong[30] that amplification similar to that obtained with

[30] See reference No. 134. Appendix a.

several stages may be secured with a single tube. Instead of feeding the voltage of the secondary coil S into the grid circuit of a second tube it is fed back into the grid circuit of the same tube so as to increase the voltage operating upon the grid. This results in an increased amplitude of the plate-current alternations which likewise being fed back into the grid circuit increases the voltage operating upon the grid, etc.

One form of the so-called feed-back circuit for rectifying and amplifying damped oscillations is shown in Fig. 145. The operation of the circuit, used as a receiving device, is the same as that described above for the case of a condenser in the grid lead. The condenser C_2 is merely to provide a path of low impedance across the phones for the high-frequency oscillations. The coils P and S constitute the feed back by means of which the oscillations in

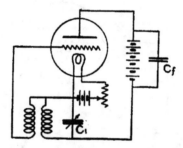

FIG. 146.—*Use of electron tube as a generator*

the tuned circuit are reinforced. The mutual inductance between S and P must be of the proper sign so that the emf fed back aids the oscillations instead of opposing them.

58. ELECTRON TUBE AS GENERATOR

Generation of Oscillations.—If the coupling between the coils P and S in Fig. 145 is continuously increased and the values of L, S, and C and the resistance of this circuit are suitable within certain limits, the emf fed back by the coil P into the oscillatory circuit at any instant will become greater than that required to just sustain the oscillations in the circuit. In this case any oscillation, however small in the circuit L, S, C, will be continuously built up in amplitude until a limit determined by the characteristics of tube and circuits is reached. In other words, the tube self-generates alternating current of a frequency determined by the natural frequency of the oscillatory circuit.

Numerous circuits have been devised to produce oscillations. Fig. 146 shows a method of connection which is suitable for producing large currents. The oscillatory circuit is in the filament-plate circuit and a coil between filament and grid. The operation of this circuit is somewhat different from that outlined above. Instead of transferring all of the energy necessary to sustain the oscillations from the plate to the grid circuit as in the preceding case, only an emf which serves as a control is here transferred. Thus, the grid circuit plays a similar part to that of the slide valve in a reciprocating engine. The path of the current flow within

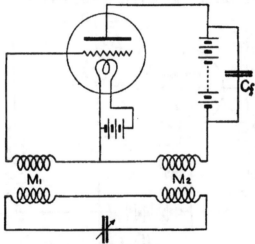

FIG. 147.—*Generating circuits in which the oscillatory circuit is inductively coupled to both the grid and plate circuits*

the tube from plate to filament may be regarded as a variable resistance, the value of which depends upon the potential of the grid. If the potential of the grid is alternating, the resistance will increase and decrease in accordance, thus throwing an alternating emf upon the oscillatory circuit in series with this resistance.

The oscillatory circuit which determines the frequency may be a separate circuit, as in Fig. 147. Here the coupling M_2 supplies the emf to reinforce the oscillations and M_1 furnishes the emf to the grid. The condenser C_f is a large fixed condenser which serves as a path of low impedance across the battery for the high-frequency alternations in the plate circuit.

In addition to the above types of circuit in which electromagnetic coupling between the plate and grid circuits is used to transfer emfs from one to the other, there are also circuits in which electrostatic coupling is utilized. This is illustrated in Fig. 148, in which the condenser C_2 serves as the coupling. The inductances L_1 and L_2 should be variable and approximately equal. C_f is a fixed condenser which serves as a path of small impedance for the high frequency around the battery. The frequency is primarily determined by the inductances L_1 and L_2 and the condenser C_2. The parallel connection of C_1 and L_1 serves as an "absorbing" circuit—that is, as C_1 is increased from a very low

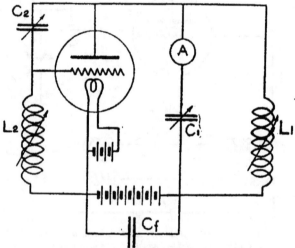

FIG. 148.—*Generating circuits in which the plate and grid circuits are electrostatically coupled*

value—the current circulating around this circuit will increase up to a certain point and may considerably exceed the current in the other portions of the circuit.

Reception of Undamped Oscillations.—If two sources, which separately furnish undamped oscillations of, say, 100 000 and 101 000 frequency, as shown in (a) and (b) of Fig. 149, act together upon the same circuit, the resultant oscillations in the circuit, obtained by adding the components, will be of the form shown in (c). The mode of adding the components is illustrated in Fig. 150. The amplitude of the combined oscillation will rise and fall, becoming a maximum when the component oscillations are in phase and a minimum when they are 180° out of phase. The beats or periodic rise and fall in amplitude occur at a rate equal to the difference in frequencies of the two oscillations. Thus, the

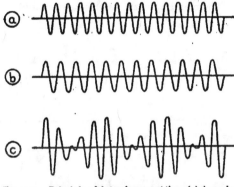

FIG. 149.—*Principle of heterodyne reception; (a) incoming oscillations, (b) oscillations produced by the tube, (c) resultant current*

beat frequency in the case assumed above would be 101 000 − 100 000 = 1000 per second. If rectified, these beats will produce a note in a telephone of like frequency. In the reception of undamped signals by this method, called the heterodyne method,

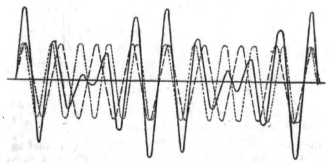

FIG. 150.—*Mode of adding component oscillations*

the incoming signals represent one component oscillation. The other oscillation is generated in the receiving apparatus and both act in the same circuit. The rectified resultant furnishes a musical note in the phones, the pitch of which can readily be altered by varying the frequency of the local source of oscilla-tions. The electron tube may serve as a convenient source of local oscillations and at the same time as an amplifier and detector of the received signals. This is called the autodyne method. Numerous circuits may be utilized to produce these results, of which that shown in Fig. 145, page 209, may serve as an illustra-tion. Incoming signals set up oscillations in the antenna. By means of the coupling between the antenna and coil L oscillations

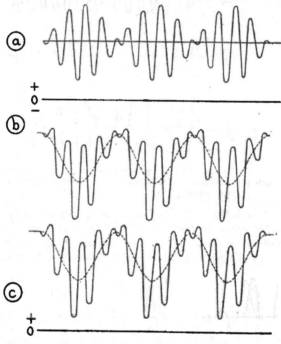

Fig. 151.—*Variations of mean grid voltage and mean plate current as beat oscillations are being produced*

of the same frequency are set up in the circuit L C, and as ex-plained above are amplified on account of the feed back between S and P. Further, the coupling between S and P is such that the tube oscillates, the frequency of these oscillations depending largely upon the constants of the circuit L C. If this latter fre-quency is adjusted to be slightly different from that of the incom-ing oscillations, beats will result and the potential of the grid will follow the beat oscillations. Just as explained before in the case of reception with a grid condenser, there will be an increased flow of negative electricity from the filament to the grid when this latter is positive and its mean potential will be lowered. Thus, as the oscillations in the beat are increasing the potential of the grid will become lower. The plate current will follow the variations in potential of the grid, reproducing the beat oscillations and decreasing in mean value as the mean potential of the grid is lowered. The curve (*a*) of Fig. 151 represents the beat oscillations in the circuit *L C*. In (*b*) is shown the oscillations of the grid potential, the mean potential being indicated by a dotted line.

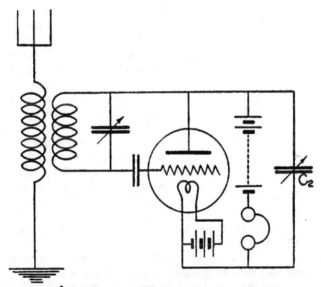

FIG. 152.—"*Ultraudion*" *circuit for receiving undamped oscillations*

In (*c*) is shown the plate current, the mean value of which is also shown by a dotted line. The telephone current will likewise correspond to this mean value and hence the note will correspond to the beat frequency. In Fig. 152 is shown the connections for the circuit used by L. De Forest, the inventor of the audion, for the reception of undamped oscillations and called the "ultraudion." The oscillatory circuit is connected between the grid and plate with a condenser in the grid lead. The variable condenser *C*, shunted across the plate battery and phones is important in the production of oscillations; in general, its value can not be increased beyond a certain point without stopping the oscillations.

By this beat method high sensitiveness and selectivity are attained in receiving. Interference is minimized because even slight differences in frequency of the waves from other sources result in notes either of different pitch or completely inaudible.

Tone Modulation of Radio Currents from Electron Tube.—In undamped wave radio transmitters, the radio-frequency currents may be modulated by the use of what may be called "tone circuits." It is then possible to take advantage of the very selective tuning obtainable with undamped waves without employing beat methods of reception. Since undamped high-frequency currents can be produced from electron tubes, it is particularly convenient to apply to these tubes devices for impressing an audible tone on the currents generated. This means that a periodic variation of the amplitude of the radio-frequency current is produced, this periodic variation being of audible frequency. Radio currents modulated in this way may be produced from electron tubes in the three ways described below.

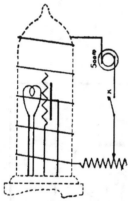

Fig. 153.—*Modulation of generated current by action of low-frequency magnetic field on electron stream*

Modulation of Electron Stream.—The electron current through the tube may be modified by placing the tube in a strong magnetic field which varies in strength with an audible period. The circuits shown in Fig. 146 or in Fig. 160 may be used. A coil wound around the electron tube is supplied with current from a 500-cycle generator, as in Fig. 153, or a direct current through the coil may be interrupted by means of a buzzer.

Modulation of Grid Potential.—Instead of modulating the current by external means, advantage may be taken of the characteristics of the tube itself. The potential of the grid with respect to the filament may be varied with a relatively slow period by means of the arrangement shown in Fig. 154, where an audio-frequency circuit is inserted in the lead to the grid of the electron tube. The circuit L_2C_2 is tuned to resonance with the L_3 circuit which is coupled loosely to it. Any of the various methods of generating alternating currents may be used for this purpose if a circuit such as L_2C_2 is inserted in the grid lead and an alternating current of an audible frequency induced in it. For

example, the L_2 circuit may be supplied from a 500-cycle generator, with L_2 equal to 50 millihenries and C_2 equal to 2 microfarads.

The audio-frequency current may, like the radio-frequency
current, be generated by an electron tube.   In this case the L_2

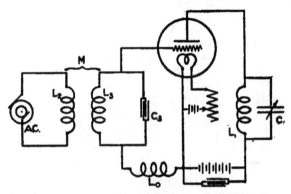

FIG. 154.—*Modulation of generated current by means of periodic changes
of grid voltage*

circuit referred to above is replaced by the oscillatory circuit of
the audio-frequency generator as in Fig. 155.   It will be noted
that the same type of circuit is used for generating audio as for
generating radio currents, it being necessary merely to provide

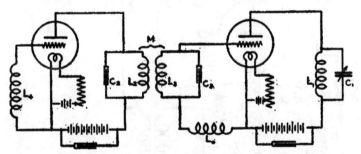

FIG. 155.—*Method of using an electron tube for producing periodic changes of grid voltage
in another tube generating radio-frequency current*

suitable values of L_4, L_2 and C_2.   There must be mutual inductance between L_4 and L_2 of Fig. 155 just as between the coils in the
grid and plate circuits of Fig. 146.   The audio and the radio
frequency generators may be operated from the same batteries.

Self-Modulating Tube.—In the methods previously described, means for modulating are provided outside of the radio-frequency tube. This, however, is not necessary, for it is possible to generate

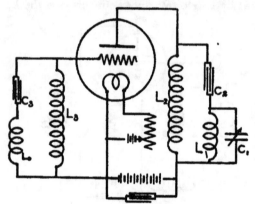

FIG. 156.—*Arrangement for producing modulated radio-frequency current by use of a single tube*

both audio and radio frequency currents simultaneously from the same tube. Two arrangements of circuits whereby this may be done are shown in Figs. 156 and 157. In these two diagrams the

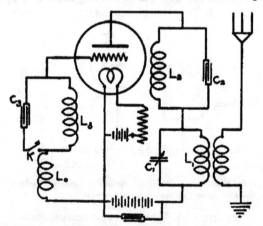

FIG. 157.—*Method of coupling tone-circuit generator to an antenna*

radio-frequency circuit is L_1C_1. The circuits L_2C_2 and L_3C_3 are of audio-frequency. There is mutual inductance between L_2 and L_3 as well as between L_0 and L_1.

FIG. 158.—*Arrangement of apparatus in the tone-circuit generator*

FIG. 159.—*Arrangement of apparatus in pliotron generator*

Any of the arrangements described above may be used to produce modulated radio-frequency currents in an antenna by coupling the antenna to the radio-frequency coil L_1 as in Fig. 157. For making signals a key may be inserted in the lead to the grid or in the connection between the filament and the B battery, or in the audio circuit L_3C_3 as shown in Fig. 157.

The pitch of the note given by these tone transmitters may be varied at will by changing the constants of the audio circuits L_2C_2 and L_3C_3. Thus, several transmitters may operate using the same wave length but having different modulating tones. The receiver, if provided with means for tuning to the audio as well as to the radio frequency, will be free from interference even by other stations using the same wave length. This method of transmitting offers the considerable advantage that the tone is a pure musical note and does not change in pitch with slight changes in the tuning of the receiving station as in beat methods of reception.

Electron Tube as Generator for Measurement Purposes.—It is desirable in generating oscillations for measurement purposes that the amplitude and frequency of the generated current shall be constant and that the set-up shall be simple and flexible. By the latter term is meant that a wide range of wave lengths may be obtained with the same apparatus.

Constancy of amplitude and frequency are easily obtained. The main requirement being steadiness in the batteries supplying the filament heating current and the electron current between plate and filament. High-frequency current, constant both in magnitude and in frequency to better than one-tenth of 1 per cent over long intervals of time, is readily obtained. When two or more tubes are operated in parallel on the same B battery changes occur in the intensity of the current furnished by one tube at the instant the second tube is put into operation or when the operation of the second tube is changed. Independent filament batteries should always be used. The circuit shown in Fig. 146 is simple and flexible. The frequency generated is ordinarily varied by changing the capacity C_1. With given coils, as the capacity is increased, there comes a point where the oscillatory current falls off and finally "breaks". It is then necessary to use coils of greater inductance in order to obtain longer wave lengths.

Another circuit similar to the above and which has shown itself to be convenient is shown in Fig. 160. Here the coils L_1

and L_2 may be wound in a single layer adjacent to each other
on the same form. Taps may be brought out on each coil so as
to use the number of turns desired. The condensers C_2 and C_3
are large fixed-value condensers which should be of low resistance.
C_1 is the tuning condenser. A tungsten lamp is introduced in
series with the B battery to protect the filament of the tube in
case of an accident. The measuring circuit may be coupled
directly to the coils L_1, L_2, or to a special coil of a few turns
inserted in series with either of these coils, preferably on the side
connected to the B battery since this point is held at constant
potential by the large capacity of battery to ground.

The B battery may be inserted directly in the lead from the
plate instead of adjacent to the filament as shown above. With

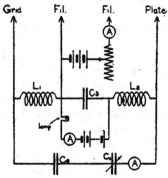

FIG. 160.—*Scheme of connections for
pliotron generator*

such connection, however, care must be taken that there is very
little capacity between the two batteries or their leads; if the
batteries or their leads are not well separated and insulated from
each other, the high-frequency current is much reduced. An
advantage of locating the B battery adjacent to the plate is that
a single continuous coil may provide all the inductances required
in the circuits. Thus, as shown in Fig. 161 below, connections
may be made to the coil LL from filament, grid, plate, condenser,
and high-frequency ammeter by movable contacts. Great lati-
tude of adjustment of the several inductances is thus allowed,
and the connections are very simply shifted from one type of
circuit to another, so that the proper connections to give maximum current for any wave length are made by simply sliding

these contacts. An advantage of the mode of drawing the circuits shown in Fig. 161 is that it brings out that the several types of connection are equivalent.

59. POULSEN ARC

Another valuable source of undamped oscillations for measurements with moderate or high power is the Poulsen arc. If, as in Fig. 162, an ordinary direct-current carbon arc in air is shunted by a circuit containing capacity and inductance in series, oscillations may be obtained in the shunt circuit. Since the oscillations obtained with this simple arc are, in general, of audible frequency, the arrangement is called the singing arc. Numerous attempts

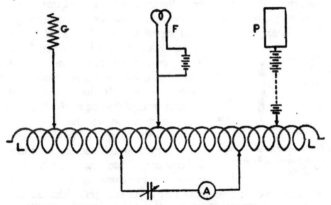

FIG. 161.—*Pliotron generator using a single coil with sliding contacts*

have been made to utilize the arc in air as a generator of high-frequency currents, but it was found that the power of the oscillations rapidly decreased with increasing frequency so that it was impossible to attain frequencies higher than about 10 000.

V. Poulsen, however, found that by modifying the arc in the following respects, high powers could be obtained at least up to moderately high radio frequencies:

1. The arc is surrounded by a hydrocarbon atmosphere such as coal gas or alcohol vapor.

2. Copper instead of carbon is substituted for the positive electrode.

Further, it is desirable to cool the copper electrode by water circulation, to rotate the carbon electrode, and (particularly for high powers) to provide a transverse magnetic field across the arc to blow it out. The source supplies several hundred volts. The action of the arc in generating oscillations is roughly the following: When no current is flowing through the gap a high voltage is required to start the arc. Immediately, however, upon starting the arc the path of the discharge is ionized and the resistance of the arc is greatly reduced; in fact, the greater the current through the arc the greater the ionization and the lower its resistance. In series with the direct-current source of supply are choke coils and regulating resistances which tend to keep the supply current constant. Suppose that the arc is suddenly extinguished and deionized. On account of the magnetic energy in the supply circuit the voltage across the arc will rise very rapidly, at the same time charging the condenser in the oscillatory circuit until a sufficient voltage is reached to strike the arc and again ionize the path of the discharge. The resistance of the arc immediately falls,

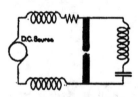

FIG. 162.—*Production of high-frequency currents by means of the Poulsen arc*

hence the condenser discharges through it, and on account of the inertia of the discharge, becomes charged again in the opposite sense. It then starts to discharge through the arc again, but in the opposite direction to the flow of current from the source. Thus, the resultant current through the arc is reduced, and when the discharge current of the condenser increases up to that of the supply, the resultant becomes zero. At this point the arc is extinguished, and, as a result of the features introduced by Poulsen, is rapidly deionized. The supply current completes the condenser discharge and again charges up the condenser to the point where the arc will strike again, and the cycle is repeated. It can readily be seen that while the discharge of the condenser is dependent upon the natural period of the oscillatory circuit, the charging depends upon such factors as arc length, constants of the supply circuit, etc., so that the period of the oscillation likewise depends upon these latter factors. Further, the voltage to which the condenser is charged, and hence the amplitude of the oscillation, depends upon the length of the arc, rapidity of deionization, etc., so that the one factor of arc length affects both the frequency and intensity of the high-frequency oscillations.

In order to obtain constancy in the oscillations such as is necessary for measuring purposes it is necessary that the arc length remain constant. When the arc is burning it tends to eat into the electrodes, and thereby increase its length, and then move to another spot with a shorter gap. This results in unsteadiness in both the frequency and amplitude of the oscillations. In some constructions the arc is caused to revolve slowly around a cylindrical electrode by means of a radial magnetic field, in others one of the electrodes is slowly revolved. In either case care must be taken to insure that the distance between the electrodes shall be constant; otherwise slow changes in frequency and intensity will result. In all cases a transverse field used to gain high power will increase the irregularities. In general, the fluctuations are minimized as the capacity in the oscillatory circuit is decreased, the wave length increased, and the supply current increased. It is practically impossible to attain reasonable steadiness in the operation at wave lengths much shorter than 1000 meters, though satisfactory operation is attainable at longer waves.

60. HIGH-FREQUENCY ALTERNATORS AND FREQUENCY TRANSFORMERS

The direct generation of high-frequency currents by means of alternators is a difficult problem; in general, very high speeds of rotation are required, and the losses in the machine from eddy currents, hysteresis and dielectric absorption are likely to be very great. However, two types of generators have been successfully evolved and have been developed to very high powers for radio transmission.

Inductor Alternator.—The first of these is of the inductor type, which has been developed by E. F. W. Alexanderson, of the General Electric Co. This machine has stationary field and armature windings and a solid steel rotor provided with slots cut at equal intervals near the circumference and filled with a nonmagnetic material. As the rotor revolves the magnetic circuit of the field is closed alternately through the nonmagnetic material filling the slots and the steel between the slots; thus the magnetic flux due to the field is alternately decreased and increased. This flux threads the armature coils, setting up an alternating emf in these. By providing the rotor with 300 slots around the circumference and driving it at a speed of 20 000 revolutions per minute a frequency of 100 000 cycles per second is attained. In a later design the frequency has been increased to 200 000 cycles per second.

It is stated that there is no difficulty in attaining a constant speed with this machine, and hence it should be of great value for measuring purposes within the range of frequencies covered. A

35601°—18——15

further extremely valuable feature is that the frequency can be determined absolutely from the speed.

Goldschmidt Alternator.—A second type of alternator is the so-called reflection type due to R. Goldschmidt. The rotor and stator are each laminated and provided with windings. The principle upon which the operation of this generator is based is as follows: If an alternator is excited with alternating current of frequency N_1, it will generate current of two frequencies $N_1 + N_2$ and $N_1 - N_2$ where N_2 is the frequency which would be generated with direct-current excitation. If $N_1 - N_2 - N$ then the frequencies would be $2N$ and o. If the current of frequency $2N$ is used to excite the field of another similar generator running at the same speed, generated frequencies of $2N + N$ and $2N - N$—that is, $3N$ and N—would result. Thus, a series of generators running at a moderately high speed could be used for generating high-frequency currents. In the Goldschmidt generator this frequency multiplication is attained in one machine. The stator is excited with direct current and current of frequency N is generated in the rotor. Since the induction of currents depends only upon the relative motion of rotor and stator we may consider that the rotor field is excited with current of frequency N and that the stator is rotating in this field. Consequently, currents of frequency $2N$ and o will be generated in the stator. The fields of these currents in turn react upon the rotor, producing in it currents of frequency $3N$ and N, and in this manner the frequency is successively stepped up, the frequencies in the rotor being odd multiples of N and those in the stator even multiples. In order that the flow of current of these frequencies may not be prevented by the reactance of the circuits, the principle of resonance is utilized and tuned circuits are provided for each frequency up to that which is to be used. The flow of current corresponding to the lower frequencies is suppressed to a great extent. For as we have seen, starting with the fundamental frequency N, after two "reflections" we again have an induced frequency N in company with $3N$. It may be shown that these two currents of frequency N will be opposite in phase and hence tend to neutralize each other. This is likewise true of the magnetic fields so that the losses due to hysteresis and eddy currents will be caused mainly by the field of the utilized frequency alone. While these machines have been developed very satisfactorily for radio transmission purposes, it is doubtful whether they could be readily utilized for measuring purposes in the laboratory since the multiplicity of tuned circuits would render frequency changes difficult.

"*Static*" *Frequency Transformers.*—Several methods of frequency multiplication have been devised which are based upon the distortion of the wave of magnetic induction in iron from that of the impressed magnetizing force. Since these frequency multipliers have no moving parts they are called static frequency transformers. The principle is well illustrated in the method of tripling the frequency, due to Joly. Fig. 163 is a typical curve showing the variation of induction in iron with the magnetizing force. As the magnetizing force is increased from zero the resultant flux of induction in the iron at first increases rather slowly, then very rapidly, and then less rapidly, again becoming almost constant at a value called the saturation value. If the magnetizing force is alternating and sinusoidal and of such an amplitude that the maximum value

comes on the steep part of the induction curve as at A, Fig. 163, the resulting alternating wave of magnetic induction will be peaked, as in b, Fig. 164. If, however, the maximum magnetizing force has a value sufficiently high to bring up the induction to the flat part of the curve where it is changing very slowly, as at B, Fig. 163, then the resulting alternating wave of induction will be flat topped, as in c, Fig. 164. The wave form b indicates that

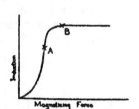

FIG. 163.—*Variation of magnetic induction in iron with magnetizing force*

there is a strong harmonic oscillation of three times the fundamental frequency impressed upon the fundamental oscillation and differing in phase from it by 180°. The wave form c likewise indicates the presence of a strong harmonic of three times the fundamental frequency but which is in phase with the fundamental. If, therefore, the two waves b and c can be combined in such a manner that the fundamental frequencies are 180° out of phase and hence neutralize each other, the harmonics of triple frequency will be in phase and will exist alone. This is illustrated in curve d which is obtained by subtracting the ordinates of the curve c from those of curve b. This method is applied by means of transformers as illustrated in Fig. 165. The alternator supplies current of the fundamental frequency f to the primaries P_1 and P_2. P_1 has few turns and P_2 many turns, so that the iron is magnetized more intensely in 2 than in 1. The two secondaries S_1 and S_2 are so wound and connected that the emf's of the fundamental frequency neutralize each other, but the triple harmonics cause a current flow in the tuned circuit of frequency $3f$. Thus with an initial frequency of 10 000 cycles per second, a frequency of 30 000 can

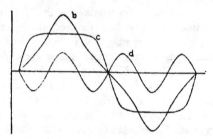

FIG. 164.—*Method of combining alternating waves of magnetic induction so as to triple the frequency*

be obtained with one transformation. This corresponds to a wave length of 10 000 meters and is suitable for long-distance transmission. Large powers may be generated.

For measuring purposes it would be possible to step up the frequency through several stages obtaining 3, 9, 27, etc., times the

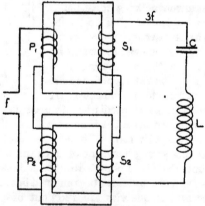

FIG. 165.—*Use of two transformers for producing frequency transformations*

fundamental frequency. This might furnish a valuable method of determining high frequencies in terms of lower frequencies, which latter can be determined absolutely from the speed and number of poles of the alternator.

61. BUZZERS

The buzzer is a very convenient source of damped oscillations for measurement purposes. Since, in general, it furnishes only very small power, it is used in conjunction with very sensitive detecting instruments. A number of different modes of connection may be used in generating oscillations with a buzzer. That shown in Fig. 166 has been found to be very satisfactory. The current from the battery *B* flows through the adjustable resistance *R*, the coils *F*, armature contact *A*, and coil L_1. When through the action of the buzzer the contact is opened, the energy due to the current in the coil L_1 is transferred to the condenser, C_1, giving it a charge. The condenser then discharges, causing a train of oscillations in the circuit

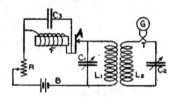

FIG. 166.—*Use of buzzer as a source of current of definite frequency*

$C_1 L_1$, the frequency of which depends upon the constants of this circuit with a small correction for the capacity added by the leads, etc., of the buzzer circuit. Thus, each break of the buzzer sets up a train of oscillations in the circuit $C_1 L_1$. The circuit $L_2 C_2$ is a measuring circuit coupled to the driving circuit $L_1 C_1$. The current in the measuring circuit may be indicated by a galvanometer and thermoelement (*T*) inserted directly in the circuit or any other sensitive device.

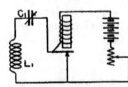

FIG. 167.—*Buzzer circuit capable of producing currents by shock excitation*

Constancy of the high-frequency current depends upon the steadiness of the buzzer action. This is obtained by using a good buzzer giving a note of high pitch, such as the Ericsson, by adjustment of the buzzer contacts and resistance *R* until the buzzer emits a clear and steady musical tone, by employing a constant battery, preferably a low-voltage storage battery to insure steady direct current, and by preventing sparking at the contact. This latter requirement is attained by sending only a moderate current through the buzzer and by using a fairly large fixed condenser C_3 across the buzzer field coils to absorb the magnetic energy stored therein which otherwise would produce a high voltage and sparking at the contact on break.

Another form of buzzer circuit which is frequently used and is capable of furnishing somewhat larger currents is shown in Fig. 167. In this case the condenser C_1 is charged to the voltage of the battery when the buzzer contact is open and discharges through L_1 when the contact is closed. A possible objection to this circuit is the presence of the buzzer contact in the oscillatory circuit.

If C_1 in the above is a fixed condenser of several microfarads capacity and L_1 a small inductance of only one or two turns, then the oscillations in the circuit $L_1 C_1$ will be very highly damped and will last for a very short time, possibly only one or two oscillations. Under these conditions an oscillatory circuit coupled to L_1 will be shocked into oscillations by what is called impact excitation, the frequency and damping of the oscillations will

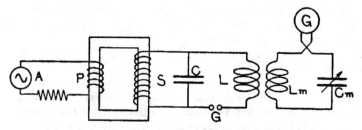

Fig. 168.—*Typical spark circuit for producing high-frequency oscillations*

be those natural to this circuit and independent of the circuit $L_1 C_1$. On this account this method of impact excitation is very useful in many measurements.

In place of the very convenient buzzer many other forms of circuit interrupters may be used, such as the vibrating wire, tuning fork, rotating and mercury interrupter.

62. THE SPARK

Certain forms of spark gap are simple and inexpensive sources of damped currents and so are often used as sources in high-frequency measurements. In some kinds of measurement it is necessary or advantageous for the oscillations to have a decrement.

Simple Spark Gap.—In Fig. 168 is shown a typical circuit for the generation of high-frequency oscillations by means of a spark discharge. The alternator supplies the low-voltage winding P of a step-up transformer. The high-voltage side S leads

to the terminals of the condenser C, across which is an inductance
L and spark gap G in series. The coil L is loosely coupled to the
measuring circuit L_mC_m (or to the antenna in transmitting).
During an alternation, as the voltage across S increases, the condenser C becomes charged up to the point where the voltage is
sufficient to jump the spark gap. The condenser then discharges
through the inductance L and the gap G. The discharge consists
of a train of oscillations of a frequency approximately corresponding to the inductance and capacity of the circuit. It is
possible to adjust the voltage of the transformer and the length
of the gap so that the discharge takes place when the voltage is
at a maximum, either positive or negative. In this case one
spark and one train of oscillations is obtained per alternation
of the supply, thus with a 60-cycle generator the spark frequency

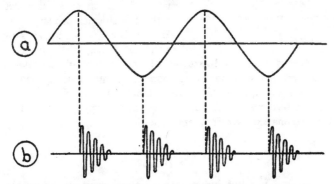

Fɪɢ. 169.—*Groups of oscillations for case of two spark discharges per cycle*

will be 120. By shortening the gap or raising the voltage several
discharges per alternation may be obtained. These are called
partial discharges and occur somewhat irregularly. The first
case is illustrated in Fig. 169. In (a) is shown the transformer
secondary voltage as it would be if the spark gap were absent
and in (b) the current oscillation in the condenser discharge. In
Fig. 170 is shown the effect of the spark gap upon the damping
of the oscillations in the high-frequency train. In a circuit with
constant resistance the amplitude would decrease exponentially
as indicated by the dash curve 1, in the case of a circuit with a
spark gap the decrease of amplitude tends to become linear, as
shown by the dash line 2. This is due to the increase in resistance
of the spark as the amplitude of the current decreases, the effect
depending upon the material of the electrodes, etc.

Use of Resonance Transformer.—A serious difficulty in the operation of the spark circuit is caused by the short-circuiting of the transformer secondary by the spark. As a result there is a heavy flow of current through the gap causing the formation of an arc which reduces the amplitude of the oscillations and destroys the electrodes. In order to eliminate this difficulty the resonance transformer is used. The alternator, transformer and secondary condenser are adjusted to make a system which is in resonance for the alternator frequency. When the condenser is short-circuited by the spark the condition of resonance is destroyed, and in effect this is equivalent to the sudden insertion of a reactance in the transformer primary. As a result, there is no heavy flow of current through the gap.

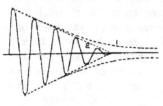

FIG. 170.—*Linear damping produced by the increase of spark resistance as the amplitude of current decreases*

The theory of the adjustment of the system of alternator, transformer, and secondary condenser to resonance is as follows. If we have a simple circuit of inductance and capacity in series across the terminals of the alternator, as in Fig. 171, the condition for resonance for a frequency f is

$$f = \frac{1}{2\pi\sqrt{L_p C_p}}.$$

where L_p is the total inductance of the circuit including the alternator. The combination of transformer and secondary condenser can be reduced to this simple case. Assuming that all the induction linked with the primary winding of the transformer also passes through the secondary turns—that is, that there is no magnetic leakage—and that the ratio of the number of secondary turns to primary turns is n, it may be shown that a capacity C_s in the secondary is

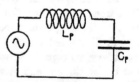

FIG. 171.—*Simple circuit equivalent to Fig. 168*

equivalent to a capacity $C_p = n^2 C_s$ in the primary. The effect of inductance in the secondary is decreased in the ratio of $1 : n^2$ when transferred back to the primary, hence inductance is inserted in the primary to tune to resonance. The total primary inductance consists of that inserted plus the inductance of the alternator and that due to transformer leakage. This latter is small in the case of a closed-core transformer. Experimentally a fairly close adjustment to resonance may readily be obtained by lowering the generator voltage until no spark passes the gap and then varying either the primary inductance or secondary condenser until the primary current or secondary voltage is a maximum. The primary inductance may conveniently consist of a solenoidal winding with an iron core that can be moved in or out to vary the inductance value.

When adjusted to resonance the voltage across the secondary condenser may rise to a value much higher than that corresponding to the voltage of the alternator and the transformer ratio. In Fig. 172 is shown the way the voltage rises with each alternation until it is sufficient to jump the spark gap discharging the condenser. The voltage then begins to rise again until the next spark takes place. The alternator voltage or spark length can be

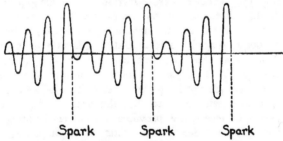

Fig. 172 —*Condenser voltage when the transformer system is adjusted to resonance with the generator*

adjusted to obtain either one spark per alternation or one spark in several alternations, as shown in the Fig. 172.

In order to obtain constant high-frequency current with a simple spark gap it is desirable to use a low spark frequency in order to prevent heating of the gap which would lead to arcing. Magnesium electrodes have been found to give the best results and to furnish oscillations most closely logarithmic in damping. Zinc is also a good material. The gap, the voltage and resonance conditions should be adjusted to give a spark of moderate and uniform frequency. The alternator must run at constant speed, otherwise the voltage and resonance conditions will vary. Under these conditions it is possible to attain high-frequency oscillations of a constancy which is probably not excelled by any other source of damped oscillations.

If the resonance transformer is not utilized, the arcing across the gap may be reduced by inserting resistance or inductance coils in the primary of the transformer and by employing an air blast to blow out the arc. Or, in place of the simple gap, a rotary gap, as shown in Fig. 173, may be utilized. Its characteristics are intermediate between those of the simple and the quenched spark gap.

Quenched Gap.—It was found by M. Wien that if a series of short spark gaps be substituted for a single long gap and a discharge passed through them, the discharge path

FIG. 173.—*Rotating spark gap*

returns much more quickly after discharge to its initial condition of high resistance. This is a result of the more rapid deionization of the gap and is called the quenching action. The quenching action is increased if the surfaces of the gaps are of silver or copper and the gap is kept cool and air-tight. In Fig. 174 is shown a cross section of a single gap showing the insulating gasket between the plates which renders the gap air-tight, the silver sparking surfaces and the flanges to provide a large cooling surface. The insulating gasket may be of paper, mica or rubber, and is about 0.2 mm thick. Its thickness is exaggerated in the figure. A number of such gaps are stacked in series and clamped together, and either the leads to the gap are provided with clips so that the number of gaps used may be varied or means are provided for short-circuiting as many of the gaps as desired. A plate of an improved quenched gap designed at the Bureau of Standards is shown in Fig. 175, facing page 323. The construction is such as to permit air circulation on both sides of each gap. This is accomplished by inverting alternate plates. The assembled quenched gap is shown in Fig. 221, page 322.

FIG. 174.—*Cross section of quenched gap plate*

While close coupling with the secondary circuit in the case of ordinary spark gaps is to be avoided, since it causes the generation of two frequencies (the so-called coupling waves, see p. 48) of which only one can be utilized, good working of the quenched gap, on the other hand, requires a fairly close coupling between the primary and secondary circuits. This secures high efficiency and still permits a single wave to be obtained. The explanation is as follows: Assume first that the primary circuit contains an ordinary spark gap, the secondary (which may be an antenna) is fairly closely coupled to the primary, and that the two circuits when separated have the same natural frequency. Due to the coupling, oscillations of two frequencies, one lower and one higher than that common to the uncoupled circuits, will result in both circuits after the discharge takes place in the primary. The combination of the two frequencies will result in beats in both circuits, the amplitude of the resultant oscillation will rise to a maximum and fall to a minimum in each circuit, being a maximum in the primary when a minimum in the secondary, and vice versa. As a result, the total energy of the oscillations (excepting that dissipated) is transferred back and forth between the two circuits. Although the current in the primary circuit may pass

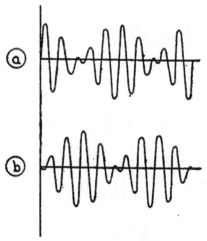

Fig. 176.—*Current in (a) primary and (b) secondary*
when using an ordinary gap

through a zero value, the rapidity of deionization of the ordinary spark gap is not sufficient to render it nonconducting in the short interval of time available and the spark reignites. The phenomena are shown in Fig. 176 where (a) represents the voltage oscillations in the primary and (b) the oscillations in the secondary.

If, on the other hand, a quenched gap is used and the coupling between the primary and secondary is favorable, it will become deionized when the primary oscillations are a minimum and thus prevent reignition. At this time all of the energy has been transferred to the secondary and, since the primary has become inoperative, this energy will be dissipated in a train of oscillations of which the frequency and damping are determined entirely by the constants of the secondary circuit. The oscillations of primary and secondary are shown in (a) and (b) of Fig. 177. In ideal operation, the time during which the primary circuit is operative will be extremely short, there will be only the one frequency, and, since the major loss of power takes place in the high-resistance primary circuit, the efficiency will be high. With poorer operation the primary circuit may remain in operation until the second or third minimum. In this case three

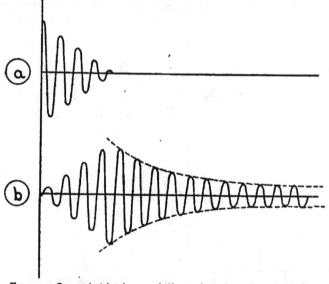

FIG. 177.—*Current in* (a) *primary and* (b) *secondary when using a quenched gap*

waves may be observed, the two coupling waves and the intermediate wave corresponding to the oscillations of the secondary by itself.

The connections for the quenched gap are similar to those for a plain spark, using a resonance transformer. Best operation is generally obtained when the inductance in the primary circuit is somewhat greater than that required for resonance. On account of the rapid quenching of the gap, the supply alternator may have a frequency of 500 cycles and adjustments made so as to obtain one spark per alternation.

PART III.—FORMULAS AND DATA

❦

CALCULATION OF CAPACITY

63. CAPACITY OF CONDENSERS

Units.—The capacities given by the following formulas are in micromicrofarads. This unit is 10^{-12} of the farad, the farad being defined as the capacity of a condenser charged to a potential of 1 volt by 1 coulomb of electricity. The micromicrofarad and the microfarad (one-millionth of a farad) are the units commonly used in radio work. Radio writers have occasionally used the cgs electrostatic unit, sometimes called the "centimeter." This unit is 1.1124 micromicrofarads.

In the formulas here given all lengths are expressed in centimeters and all areas in square centimeters. The constants given are correct[31] to 0.1 per cent.

PARALLEL PLATE CONDENSER

Let S = surface area of one plate

τ = thickness of the dielectric

K = dielectric constant ($K = 1$ for air, and for most ordinary substances lies between 1 and 10).

$$C = 0.0885 K \frac{S}{\tau} \text{ micromicrofarads.} \qquad (110)$$

If, instead of a single pair of metal plates, there are N similar plates with dielectric between, alternate plates being connected in parallel,

$$C = 0.0885 K \frac{(N-1)S}{\tau} \qquad (111)$$

In these formulas no allowance is made for the curving of the lines of force at the edges of the plates; the effect is negligible when τ is very small compared with S.

---

[31] The constants given in the formulas are correct for absolute units. To reduce to international units the values in absolute units should be multiplied by 1.00052. This difference need not be considered when calculations correct to 1 part in 1000 only are required.

VARIABLE CONDENSER WITH SEMICIRCULAR PLATES

Let N = total number of parallel plates

  r_1 = outside radius of the plates

  r_2 = inner radius of plates

  τ = thickness of dielectric

  K = dielectric constant

Then, for the position of maximum capacity (movable plates between the fixed plates),

$$C = 0.1390 K \frac{(N-1)(r_1^2 - r_2^2)}{\tau} \tag{112}$$

This formula does not take into account the effect of the edges of the plates, but as the capacity is also affected by the containing case it will not generally be worth while to take the edge effect into account.

Formula (112) gives the maximum capacity between the plates with this form of condenser. As the movable plates are rotated the capacity decreases, and ordinarily the decrease in capacity is proportional to the angle through which the plates are rotated.

ISOLATED DISK OF NEGLIGIBLE THICKNESS

Let d = diameter of the disk

then
$$C = 0.354 d \tag{113}$$

ISOLATED SPHERE

Let d = diameter of the sphere

then
$$C = 0.556\, d \tag{114}$$

TWO CONCENTRIC SPHERES

Let r_1 = inner radius of outside sphere

  r_2 = radius of inside sphere

  K = dielectric constant of material between the spheres

$$C = 1.112 K \frac{r_1 r_2}{r_1 - r_2} \tag{115}$$

TWO COAXIAL CYLINDERS

Let r_1 = radius of outer cylinder

  r_2 = radius of inner cylinder

  K = dielectric constant of material between the cylinders

  l = length of each cylinder

$$C = \frac{0.2416 l}{\log_{10} \frac{r_1}{r_2}} \tag{116}$$

This formula makes no allowance for the difference in density of the charge as the ends of the cylinders are approached.

64. CAPACITY OF WIRES AND ANTENNAS.

SINGLE LONG WIRE PARALLEL TO THE GROUND

For a single wire of length l and diameter d, suspended at a height h above the ground, the capacity is

$$C = \frac{0.2416 l}{\log_{10} \frac{4h}{d} + \log_{10} \left\{ \frac{l/2 + \sqrt{l^2/4 + d^2/4}}{l/2 + \sqrt{l^2/4 + 4h^2}} \right\}} \tag{117}$$

Usually the diameter d may be neglected in comparison with the length l, and the following equations are convenient for numerical computations.

For $\frac{4h}{l} \lessgtr 1$,

$$C = \frac{0.2416 l}{\log_{10} \frac{4h}{d} - k_1} \tag{118}$$

For $\frac{l}{4h} \lessgtr 1$,

$$C = \frac{0.2416 l}{\log_{10} \frac{2l}{d} - k_2} \tag{119}$$

in which the quantities

$$k_1 = \log_{10} \left\{ \frac{1 + \sqrt{1 + \left(\frac{4h}{l}\right)^2}}{2} \right\}$$

and

$$k_2 = \log_{10} \left\{ \frac{l}{4h} + \sqrt{1 + \left(\frac{l}{4h}\right)^2} \right\}$$

may be interpolated from Table 6, page 242.

These formulas assume a uniform distribution of charge from point to point of the wire.

VERTICAL WIRE

Formula (119), omitting the k_2 in the denominator, is sometimes used to calculate the capacity of a vertical wire. It applies accurately only when h is large compared with l, and gives very rough values for a vertical single-wire antenna, the lower end of which is connected to apparatus at least several meters above the ground.

CAPACITY BETWEEN TWO HORIZONTAL PARALLEL WIRES AT THE SAME HEIGHT

Let $d =$ the diameter of cross section of the wires
$\quad l =$ length of each wire
$\quad h =$ the height of the wires above the earth
$\quad D =$ distance between centers of the wires.

The capacity is defined as the quotient of the charge on one wire, divided by the difference in potential of the two wires, when the potential of one wire is as much positive as the other is negative.

$$C = \frac{0.1208\,l}{\log_{10}\left\{\dfrac{l/2+\sqrt{l^2/4+d^2/4}}{l/2+\sqrt{l^2/4+4h^2}}\cdot\dfrac{4h}{d}\right\}-\log_{10}\left\{\dfrac{l/2+\sqrt{l^2/4+D^2}}{l/2+\sqrt{l^2/4+D^2+4h^2}}\cdot\dfrac{\sqrt{D^2+4h^2}}{D}\right\}} \tag{120}$$

In most cases d/l and D/l may be neglected in comparison with unity, and we may write

$$C = \frac{0.1208\,l}{\log_{10}\dfrac{2D}{d}-\dfrac{D^2}{8h^2}} \tag{121}$$

TWO PARALLEL WIRES, ONE ABOVE THE OTHER

For the case of one wire placed vertically above the other, the formula (121) may usually be used, taking for the value of h the mean height of the wires, $\dfrac{h_1+h_2}{2}$. The potential of one wire is assumed to be as much positive as the other is negative.

CAPACITY OF TWO PARALLEL WIRES JOINED TOGETHER

Let $l =$ the length of each wire
$\quad D =$ distance between centers
$\quad h =$ their height above the earth
$\quad d =$ diameter of cross section.

The wires are supposed to be parallel to each other and to lie in a horizontal plane. They are joined together so that they are at the same potential. The capacity is defined as the quotient of the sum of their charges by the potential above the earth.

$$C = \frac{0.4831\,l}{\log_{10}\left\{\dfrac{l/2+\sqrt{l^2/4+d^2/4}}{l/2+\sqrt{l^2/4+4h^2}}\cdot\dfrac{4h}{d}\right\}+\log_{10}\left\{\dfrac{l/2+\sqrt{l^2/4+D^2}}{l/2+\sqrt{l^2/4+D^2+4h^2}}\cdot\dfrac{\sqrt{4h^2+D^2}}{D}\right\}} \tag{122}$$

which, in those cases where d^2/l^2 and $\left(\dfrac{D}{2h}\right)^2$ may be neglected in comparison with unity, may be written in the following forms:

For $\frac{4h}{l} \lessgtr 1$,

$$C = \frac{0.4831 \ l}{\log_{10} \frac{4h}{d} + \log_{10} \frac{2h}{D} - 2k_1} \tag{123}$$

For $\frac{l}{4h} \lessgtr 1$,

$$C = \frac{0.4831 \ l}{\log_{10} \frac{2l}{d} + \log_{10} \frac{l}{D} - 2k_2} \tag{124}$$

The quantities k_1 and k_2 are the same as in (118) and (119) and may be obtained from Table 6, page 242.

These formulas assume a uniform distribution of charge along the wire. (See p. 239.)

CAPACITY OF A NUMBER OF HORIZONTAL WIRES IN PARALLEL

This case is of importance in the calculation of the capacity of certain forms of antenna. The wires are supposed to be joined together, and thus all are at the same potential. Their capacity in parallel is then defined as the quotient of the sum of all their charges by their common potential.

An expression for this case as accurate as the preceding formula (120) for two wires would be very complicated. The following simpler solution is nearly as accurate, and in view of the disturbing effect of trees, houses, and other like objects on the capacity of an antenna, will suffice for ordinary purposes of design.

Let n = number of wires in parallel

D = spacing of wires in parallel, measured between centers

d = diameter of wire

h = height of the wires above the ground

l = length of each wire.

Then if the potential coefficients be calculated as follows:

$$\left. \begin{array}{l} p_{11} = 4.605 \left[\log_{10} \frac{4h}{d} - k \right] \\[2mm] p_{12} = 4.605 \left[\log_{10} \frac{2h}{D} - k_1 \right] \end{array} \right\} \text{for} \frac{4h}{l} \lessgtr 1, \tag{125}$$

or,

$$\left. \begin{array}{l} p_{11} = 4.605 \left[\log_{10} \frac{2l}{d} - k_2 \right] \\[2mm] p_{12} = 4.605 \left[\log_{10} \frac{l}{D} - k_2 \right] \end{array} \right\} \text{for} \frac{l}{4h} \lessgtr 1, \tag{126}$$

85601°—18——16

the approximate capacity of the n wires in parallel will be

$$C = 1.112l + \left[\frac{p_{11} + (n-1)p_{12}}{n} - k\right] \qquad (127)$$

the quantities k, k_1 and k_2 being obtained from Tables 6 and 7, page 242.

Example.—To find the capacity of an antenna of 10 wires 0.16 inch in diameter, in parallel, each wire 110 feet long, the spacing between the wires being 2 feet and their height above the ground 80 feet.

For this case $4h/l = \frac{320}{110}$ or $l/4h = 0.344$ and Table 6 gives k_2 = 0.146.

$$2l/d = \frac{2 \times 12 \times 110}{0.16} = 16500, \quad \log_{10}\frac{2l}{d} = 4.2175$$

$$l/D = \frac{110}{2} = 55 \qquad\qquad \log_{10}l/D = 1.7404$$

$$\therefore \ p_{11} = 4.605 \ [4.218 - 0.146] = 18.75$$
$$p_{12} = 4.605 \ [1.740 - 0.146] = 7.340$$

and from formula (127) and Table 7 the capacity is, reducing the length of the wires to cm

$$C = (1.112 \times 110 \times 30.5) + \left[\frac{18.75 + 9(7.340)}{10} - 2.05\right]$$
$$= 584 \ \mu\mu f = 0.000584 \ \mu f.$$

Example.—A second antenna of 10 wires, 3/32 inch diameter, 155 feet long, spaced 2.5 feet apart, and stretched at a distance of 64 feet from the earth.

For this case $l/4h = \frac{155}{256} = 0.606$, $\qquad k_2 = 0.249$

$$2l/d \qquad = 39680, \quad \log_{10}\frac{2l}{d} = 4.5986$$

$$l/D \qquad = 62, \quad \log_{10}l/D = 1.7924$$

$$p_{11} = 20.04, \ p_{12} = 7.11, \quad \frac{p_{11} + 9p_{12}}{10} - 2.05 = 6.35$$

$$C = \frac{1.112 \times 155 \times 30.5}{6.35} = 0.000829 \ \mu f.$$

If the length of the antenna had been 500 feet, with the height unchanged, then $\frac{4h}{l} = \frac{256}{500} = 0.512$, $k_1 = 0.026$, $\log_{10} \frac{4h}{d} = 4.5154$, $\log_{10} \frac{2h}{D} = 1.7093$; by (125) $p_{11} = 20.67$, $p_{12} = 7.75$, $k = 2.05$,

$$C = \frac{1.112 \times 500 \times 30.5}{6.99} = 0.002426 \; \mu f.$$

65. TABLES FOR CAPACITY CALCULATIONS

TABLE 5.—For Converting Common Logarithms Into Natural Logarithms

Common	Natural	Common	Natural	Common	Natural	Common	Natural
0	0.0000	25.0	57.565	50.0	115.129	75.0	172.694
1.0	2.3026	26.0	59.867	51.0	117.432	76.0	174.996
2.0	4.6052	27.0	62.170	52.0	119.734	77.0	177.299
3.0	6.9078	28.0	64.472	53.0	122.037	78.0	179.608
4.0	9.2103	29.0	66.775	54.0	124.340	79.0	181.904
5.0	11.513	30.0	69.078	55.0	126.642	80.0	184.207
6.0	13.816	31.0	71.380	56.0	128.945	81.0	186.509
7.0	16.118	32.0	73.683	57.0	131.247	82.0	188.812
8.0	18.421	33.0	75.985	58.0	133.550	83.0	191.115
9.0	20.723	34.0	78.288	59.0	135.853	84.0	193.417
10.0	23.026	35.0	80.590	60.0	138.155	85.0	195.720
11.0	25.328	36.0	82.893	61.0	140.458	86.0	198.022
12.0	27.631	37.0	85.196	62.0	142.760	87.0	200.325
13.0	29.934	38.0	87.498	63.0	145.063	88.0	202.627
14.0	32.236	39.0	89.801	64.0	147.365	89.0	204.930
15.0	34.539	40.0	92.103	65.0	149.668	90.0	207.233
16.0	36.841	41.0	94.406	66.0	151.971	91.0	209.535
17.0	39.144	42.0	96.709	67.0	154.273	92.0	211.838
18.0	41.447	43.0	99.011	68.0	156.576	93.0	214.140
19.0	43.749	44.0	101.314	69.0	158.878	94.0	216.443
20.0	46.052	45.0	103.616	70.0	161.181	95.0	218.746
21.0	48.354	46.0	105.919	71.0	163.484	96.0	221.048
22.0	50.657	47.0	108.221	72.0	165.786	97.0	223.351
23.0	52.959	48.0	110.524	73.0	168.089	98.0	225.653
24.0	55.262	49.0	112.827	74.0	170.391	99.0	227.956
						100.0	230.259

The table is carried out to a higher precision than the formulas, e. g., 2.3026 is abbreviated to 2.303 in the formulas.

Examples.—To illustrate the use of such a table, suppose we wish to find the natural logarithm of 37.48. The common logarithm of 37.48 is 1.57380.

If we denote the number 2.3026 by M, then from the table

$$1.5 \quad M = 3.4539$$
$$.073 \quad M = .1681$$
$$.00080 \; M = .0018$$

$$3.6238 = \log_e 37.48$$

To find the natural logarithm of 0.00748: The common logarithm is 3.87390, which may be written 0.87390−3. Entering the table we find

$$0.87 \quad M = 2.00325 \quad -3 \; M = -6.9078$$
$$.0039 \; M = .00898$$

$$\text{sum} \quad 2.0122$$
$$-6.9078$$

$$-4.8956 = \text{natural log of } 0.00748$$

TABLE 6.—For Use in Connection with Formulas (118), (119), (123), (124), (125), and (126)

4h/l	k₁	1/4h	k₂	4h/l	k₁	1/4h	k₂
0	0	0	0	0.6	0.035	0.6	0.247
0.1	0.001	0.1	0.043	.7	.045	.7	.283
.2	.004	.2	.086	.8	.057	.8	.318
.3	.009	.3	.128	.9	.069	.9	.351
.4	.016	.4	.169	1.0	.082	1.0	.383
.5	.025	.5	.209				

TABLE 7.—Values of k in Formulas (127) and (146)

n	k	n	k	n	k	n	k
2	0	6	1.18	11	2.22	16	2.85
3	0.308	7	1.43	12	2.37	17	2.95
4	.621	8	1.66	13	2.51	18	3.04
5	.906	9	1.86	14	2.63	19	3.14
		10	2.05	15	2.74	20	3.24

CALCULATION OF INDUCTANCE

66. GENERAL

In this section are given formulas for the calculation of self and mutual inductance in the more common circuits met with in practice. The attempt is here made, not to present all the formulas available for this purpose, but rather the minimum number required, and to attain an accuracy of about one part in a thousand. So far as has seemed practicable, tables have been prepared to facilitate numerical calculations. In some cases, to render interpolation more certain, the values in the tables are carried out to one more significant figure than is necessary. In such instances, after having obtained the required quantity by interpolation from a table, the superfluous figure may be dropped. In all the tables the intervals for which the desired quantities are tabulated are taken small enough to render the consideration of second differences in interpolation unnecessary.

Most of the formulas given are for low frequencies, this fact being indicated by the subscript zero, thus L_0, M_0. The high-frequency formulas are given where such are known. Fortunately it is possible by proper design to render unimportant the change of inductance with frequency, except in cases where extremely high precision is required.

The usual unit of inductance used in radio work is the microhenry, which is one millionth of the international henry.[22] The

---

[22] The constants in the formulas for inductance given here refer to absolute units. To reduce to international units multiply by 0.99948. Since, however, an accuracy of the order of only one part in a thousand is sought here, it will not be necessary to take this difference into account.

henry is defined as the inductance "in a circuit when the electromotive force induced in this circuit is one international volt, while
the inducing current varies at the rate of one ampere per second."
1 henry = 1000 millihenries = 10^6 microhenries = 10^9 *cgs* electromagnetic units.

In the following formulas lengths and other dimensions are
expressed in centimeters, unless otherwise stipulated, and the
inductance calculated will be in microhenries.

Logarithms are given, either to the natural base ϵ or to the
base 10, as indicated. The labor involved in the multiplication
of common logarithms by the factor 2.303 to reduce to the corresponding natural logarithms will be very materially reduced by
the employment of the multiplication table, Table 5, page 124, 241
which is an abridgement of the table for this purpose usually given
in collections of logarithms.

All of these formulas assume that there is no iron in the vicinity
of the conductor or circuit of which the inductance is to be calculated. Thus, the formulas here given can not be used to calculate
the inductance of electromagnets.

A much more complete collection of inductance formulas with
numerical examples is given in the Bulletin of the Bureau of
Standards, 8, pages 1–237; 1912; also known as Scientific Paper
No. 169.

67. SELF-INDUCTANCE OF WIRES AND ANTENNAS

STRAIGHT, ROUND WIRE

If l = length of wire
   d = diameter of cross section
   μ = permeability of the material of the wire

$$L_0 = 0.002l \left[\log_\epsilon \frac{4l}{d} - 1 + \frac{\mu}{4} \right] \text{ microhenries} \qquad (128)$$

$$= 0.002l \left[2.303 \log_{10} \frac{4l}{d} - 1 + \frac{\mu}{4} \right] \text{ microhenries} \qquad (129)$$

For all except iron wires this becomes

$$L_0 = 0.002l \left[2.303 \log_{10} \frac{4l}{d} - 0.75 \right] \qquad (130)$$

For wires whose length is less than about 1000 times the diameter
of the cross section $\left(\frac{2l}{d} < 1000 \right)$, the term $\frac{d}{2l}$ should be added inside
the brackets. These formulas give merely the self-inductance

of one conductor. If the return conductor is not far away, the mutual inductances have to be taken into account (see formulas (134) and (136)).

As the frequency of the current increases, the inductance diminishes, and approaches the limiting value.

$$L_\infty = 0.002l \left[2.303 \log_{10} \frac{4l}{d} - 1 \right] \tag{131}$$

which holds for infinite frequency.

The general formula for the inductance at any frequency is

$$L = 0.002l \left[2.303 \log_{10} \frac{4l}{d} - 1 + \mu\delta \right] \tag{132}$$

where δ is a quantity given in Table 8, page 282, as a function of x where

$$x = 0.1405d \sqrt{\frac{\mu f}{\rho}} \tag{133}$$

$f =$ frequency.
$\rho =$ volume resistivity of wire in microhm-centimeters
$\rho_0 =$ same for copper
$\mu = 1$ for all except iron wires.

For copper at 20° C, $x_0 = 0.1071\ d\ \sqrt{f}$.

The value a_c of x for a copper wire 0.1 cm in diameter at different frequencies may be obtained from Table 19, page 311. For a copper wire d cm in diameter $x_c = 10\ d\ a_c$ and for a wire of some other material $x = 10\ d\ a_c \sqrt{\mu \dfrac{\rho_c}{\rho}}$.

The total change in inductance when the frequency of the current is raised from zero to infinity is a function of the ratio of the length of the wire to the diameter of the cross section. Thus, the decrease in inductance of a wire whose length is 25 times the diameter is 6 per cent at infinite frequency; and for a wire 100 000 times as long as its diameter, 2 per cent.

Example.—For a copper wire of length 200 cm and diameter 0.25 cm at a wave length of 600 meters, that is $f = 500\ 000$, the value of x is 18.93, and from Table 8, $\delta = 0.037$.

$$\mu = 1, \quad \frac{4l}{d} = 3200, \quad \log_{10} 3200 = 3.51851$$

(From Table 5)

$$\log_e 3200 = 8.0590$$
$$414$$
$$12$$
$$\overline{8.1016}$$

For zero frequency

$$L_0 = 0.4 \, [8.102 - 1 + 0.25] = 2.941 \text{ microhenry}$$

For $f = 500\,000$

$$L = 0.4 \, [8.102 - 1 + 0.037] = 2.856 \text{ microhenry}$$

a difference of 2.9 per cent out of a possible 3.4 per cent.

For an iron wire of the same length and diameter, assuming a resistivity 7 times as great as that of copper, and a permeability of 100, the value of x is $\sqrt{\dfrac{100}{7}}$ times as great as for the copper wire, or 71.5, and for this value of x,

$$\delta = 0.010 \text{ (Table 8)}$$

$$L_0 = 0.4 \, [32.10] = 12.84 \, \mu h$$

$$L = 0.4 \, [8.102] = 3.24 \, \mu h \text{ at } 500\,000 \text{ cycles.}$$

The limiting value is $L_\infty = 2.84 \, \mu h$.

TWO PARALLEL, ROUND WIRES—RETURN CIRCUIT

In this case the current is supposed to flow in opposite directions in two parallel wires each of length l and diameter d. Denoting by D the distance from the center of one wire to the center of the other,

$$L = 0.004 \, l \left[2.303 \log_{10} \frac{D}{d} - \frac{D}{l} + \mu\delta \right] \tag{134}$$

The permeability of the wires being μ, and δ being obtained from (133) and Table 8, page 282. For low frequency $\delta = 0.25$. This formula neglects the inductance of the connecting wires between the two main wires. If these are not of negligible length, their inductances may be calculated by (132) and added to the result obtained by (134), or else the whole circuit may be treated by the formula (138) for the rectangle below.

STRAIGHT RECTANGULAR BAR

Let $l =$ length of bar.

b, $c =$ sides of the rectangular section.

$$L_0 = 0.002\, l \left[2.303 \log_{10} \frac{2l}{b+c} + 0.5 + 0.2235 \frac{(b+c)}{l} \right] \qquad (135)$$

The last term may be neglected for values of l greater than about 50 times $(b+c)$.

The permeability of the wire is here assumed as unity.

RETURN CIRCUIT OF RECTANGULAR WIRES

If the wires are supposed to be of the same cross section, b by c, and length l, and of permeability unity, and the distance between their centers is D,

$$L_0 = 0.004\, l \left[2.303 \log_{10} \frac{D}{b+c} + \frac{3}{2} - \frac{D}{l} + 0.2235 \frac{(b+c)}{l} \right] \qquad (136)$$

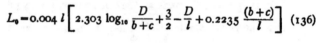

FIG. 178.—*The two conductors of a return circuit of rectangular wires*

For wires of different sizes, the inductance is given by $L_0 = L_1 + L_2 - 2M$ in which the inductances L_1 and L_2 of the individual wires are to be calculated by (135), and their mutual inductance M by (174) below.

SQUARE OF ROUND WIRE

If a is the length of one side of the square and the wire is of circular cross section of diameter d, the permeability of the wire being μ,

$$L = 0.008\, a \left[2.303 \log_{10} \frac{2a}{d} + \frac{d}{2a} - 0.774 + \mu\delta \right] \qquad (137)$$

in which δ may be obtained from Table 8 as a function of the argument x given in formula (133). The value of δ for low frequency is 0.25, and for infinite frequency is 0.

RECTANGLE OF ROUND WIRE

Let the sides of the rectangle be a and a_1, the diagonal $g = \sqrt{a^2 + a_1^2}$ and $d =$ diameter of the cross section of the wire. Then the inductance at any frequency is

$$L = 0.00921 \left[(a + a_1) \log_{10} \frac{4aa_1}{d} - a \log_{10} (a + g) - a_1 \log_{10} (a_1 + g) \right]$$
$$+ 0.004 \left[\mu\delta (a + a_1) + 2 (g + d/2) - 2 (a + a_1) \right] \qquad (138)$$

The quantity δ is obtained by use of (133) and Table 8. Its value for zero frequency is 0.25, and is 0 for infinite frequency.

RECTANGLE OF RECTANGULAR-SECTION WIRE

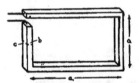

Fig. 179.—*Rectangle of rectangular wire*

Assuming the dimensions of the section of the wire to be b and c, and the sides of the rectangle a and a_1, then for nonmagnetic material the inductance at low frequency is

$$L_0 = 0.00921 \left[(a + a_1) \log_{10} \frac{2aa_1}{b + c} - a \log_{10} (a + g) - a_1 \log_{10} (a_1 + g) \right]$$
$$+ 0.004 \left[2g - \frac{a + a_1}{2} + 0.447 (b + c) \right] \qquad (139)$$

where $g = \sqrt{a^2 + a_1^2}$.

INDUCTANCE OF GROUNDED HORIZONTAL WIRE

If we have a wire placed horizontally with the earth, which acts as the return for the current, the self-inductance of the wire is given by the following formula, in which

$l =$ length of the wire
$h =$ height above ground
$d =$ diameter of the wire
$\mu =$ permeability of the wire
$\delta =$ constant given in Table 8, to take account of the effect of frequency (see p. 282).

$$L = 0.004605\, l \left[\log_{10} \frac{4h}{d} + \log_{10} \left\{ \frac{l + \sqrt{l^2 + d^2/4}}{l + \sqrt{l^2 + 4h^2}} \right\} \right]$$
$$+ 0.002 \left[\sqrt{l^2 + 4h^2} - \sqrt{l^2 + d^2/4} + \mu\delta - 2h + \frac{d}{2} \right] \qquad (140)$$

which, neglecting $\frac{d}{l}$, as may be done in all practical cases, may be written in the following forms convenient for calculation:

For $\frac{2h}{l} \lessgtr 1$,

$$L = 0.002 \ l \left[2.3026 \ \log_{10} \frac{4h}{d} - P + \mu\delta \right] \tag{141}$$

and for $\frac{l}{2h} \lessgtr 1$,

$$L = 0.002 \ l \left[2.3026 \ \log_{10} \frac{4l}{d} - Q + \mu\delta \right] \tag{142}$$

the values of P and Q being obtained by interpolation from Table 9.

Mutual Inductance of Two Parallel Grounded Wires.—The two wires are assumed to be stretched horizontally, with both ends grounded, the earth forming the return circuit.

Let l = length of each wire

   d = diameter of wire

   D = distance between centers of the wires

   h = height above the earth

Then

$$M = 0.004605 \ l \left[\log_{10} \frac{\sqrt{4h^2 + D^2}}{D} + \log_{10} \left\{ \frac{l + \sqrt{l^2 + D^2}}{l + \sqrt{l^2 + D^2 + 4h^2}} \right\} \right]$$
$$+ 0.002 \ l \ [\sqrt{l^2 + D^2 + 4h^2} - \sqrt{l^2 + D^2} + D - \sqrt{D^2 + 4h^2}] \tag{143}$$

which, if we neglect $\frac{D^2}{l^2}$ and $\left(\frac{D}{2h}\right)^2$ may be expressed in the following forms:

For $\frac{2h}{l} \lessgtr 1$,

$$M = 0.002 \ l \left[2.3026 \ \log_{10} \frac{2h}{D} - P + \frac{D}{l} \right] \tag{144}$$

and for $\frac{l}{2h} \lessgtr 1$,

$$M = 0.002 \ l \left[2.3026 \ \log_{10} \frac{2l}{D} - Q + \frac{D}{l} \right] \tag{145}$$

the values of the quantities P and Q being obtained by interpolation from Table 9.

INDUCTANCE OF GROUNDED WIRES IN PARALLEL

The expressions for the inductance of n grounded wires in parallel involve the inductances of the single wires and the mutual inductances between the wires. Even in the case that the wires are all alike and evenly spaced, these expressions are very complicated.

The following approximate equation, which neglects the resistances of wires, is capable of giving results accurate to perhaps 1 per cent, for n wires of the same diameter evenly spaced.

Calculate by equations (141), (142), (144), or (145) the inductance L_1 per unit length of a single wire and the mutual inductance M_1 per unit length of any two adjacent wires using, of course, the actual length in the calculation of the ratios $\frac{2h}{l}, \frac{2l}{d}$, etc. Then

$$L = l \left[\frac{L_1 + (n-1)\,M_1}{n} - 0.001\,k \right] \tag{146}$$

in which n is the number of wires in parallel and k is a function of n tabulated in Table 7, page 242.

Example.—An antenna of 10 wires in parallel, each wire 155 feet long and $\frac{1}{12}$ inch in diameter, spaced 2.5 feet apart, and suspended at a height of 64 feet above the earth. Find the inductance at 100 000 cycles per second.

We have here $\frac{2h}{l} = \frac{128}{155} = 0.826$, and using this as argument in Table 9, $P = 0.6671$.

From (133) $x = 8.07$, and thence from Table 8, $\delta = 0.087$.

$$\frac{4h}{d} = 256 \times 12 \times \frac{32}{3} = 32\ 768, \quad \log_{10}\frac{4h}{d} = 4.515$$

$$\frac{2h}{D} = \frac{128}{2.5} = 51.2 \qquad \log_{10}\frac{2h}{D} = 1.709$$

Then, from formulas (141) and (144)

$$L_1 = 0.002[4.515 \times 2.3026 - 0.667 + 0.087]$$
$$= 0.01963\ \mu h \text{ per cm}$$

$$M_1 = 0.002[1.709 \times 2.3026 - 0.667 + 0.016]$$
$$= 0.006568\ \mu h \text{ per cm.}$$

From Table 7 we find for $n = 10$, $k = 2.05$, so that the inductance as calculated by (146) is

$$L = 155 \times 30.5 \left[\frac{0.01963 + 9(0.006568)}{10} - 0.00205 \right]$$
$$= 4727 \, [0.00582] = 27.4 \, \mu h.$$

CIRCULAR RING OF CIRCULAR SECTION

If a = mean radius of ring

d = diameter of wire, the inductance at any frequency is, except for values of $\frac{d}{2a} > 0.2$,

$$L = 0.01257 \, a \left\{ 2.303 \log_{10} \frac{16a}{d} - 2 + \mu\delta \right\} \qquad (147)$$

in which δ will be obtained from (133) and Table 8, page 282. Its value for zero frequency is 0.25.

TUBE BENT INTO A CIRCLE

Let the inner and outer diameters of the annular cross section of the tube be d_1 and d_2, respectively, and the mean radius of the circle a, then neglecting $\frac{d_1^2}{a^2}$ and $\frac{d_2^2}{a^2}$

$$L_0 = 0.01257 \, a \left[2.303 \log_{10} \frac{16a}{d_2} - 1.75 - \frac{d_1^2}{2(d_2^2 - d_1^2)} \right.$$
$$\left. + 2.303 \, \frac{d_1^4}{(d_2^2 - d_1^2)^2} \log_{10} \frac{d_2}{d_1} \right] \qquad (148)$$

For infinite frequency this becomes

$$L_\infty = 0.01257 \, a \left[2.303 \log_{10} \frac{16a}{d_2} - 2 \right] \qquad (149)$$

68. SELF-INDUCTANCE OF COILS

CIRCULAR COIL OF CIRCULAR CROSS SECTION

For a coil of n fine wires wound with the mean radius of the turns equal to a, the area of cross section of the winding being a circle of diameter d,

$$L_0 = 0.01257 \, an^2 \left\{ 2.303 \log_{10} \frac{16a}{d} - 1.75 \right\} \qquad (150)$$

This neglects the space occupied by the insulation between the wires.

TORUS WITH SINGLE-LAYER WINDING

A torus is a ring of circular cross section (doughnut shape).

Let R = distance from axis to center of cross section of the winding

a = radius of the turns of the winding

n = number of turns of the winding

$$L_0 = 0.01257 \, n^2 \left[R - \sqrt{R^2 - a^2} \right] \tag{151}$$

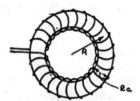

Fig. 180.—*Torus of single layer winding*

TOROIDAL COIL OF RECTANGULAR CROSS SECTION WITH SINGLE-LAYER WINDING

A coil of this shape might also be called a circular solenoid of rectangular section.

Let r_1 = inner radius of toroid (distance from the axis to inside of winding)

r_2 = outer radius of toroid (distance from axis to outside of winding)

h = axial depth of toroid.

Then $L_0 = 0.004606 \, n^2 \, h \, \log_{10} \dfrac{r_2}{r_1}$. $\tag{152}$

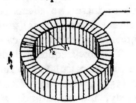

Fig. 181.—*Toroidal coil of rectangular section with single layer winding*

The value so computed is strictly correct only for an infinitely thin winding. For a winding of actual wires a correction may be calculated as shown in Bulletin, Bureau of Standards, **8**, page 125; 1912. The correction is, however, very small.

SINGLE-LAYER COIL OR SOLENOID

An approximate value is given by

$$L_s = \frac{0.03948 \ a^2 n^2}{b} K \tag{153}$$

where n = number of turns of the winding, a = radius of the coil, measured from the axis to the center of any wire, b = length of coil = n times the distance between centers of turns, and K is a function of $\frac{2a}{b}$ and is given in Table 10, page 283, which was calculated by Nagaoka. (See Bulletin, Bureau of Standards, 8, p. 224, 1912.) For a coil very long in comparison with its diameter, $K = 1$.

Formula (153) takes no account of the shape or size of the cross section of the wire. Formulas are given below for more accurate calculation of the low-frequency inductance. The inductance at high frequency can not generally be calculated with great accuracy. Formulas which take account of the skin effect, or change of current distribution with frequency, have been developed. The change is very small when the coil is wound with suitably stranded wire. The inductance at high frequencies depends, however, also on the capacity of the coil, which is generally not calculable. If the capacity is known, from measurements or otherwise, its effect upon the inductance can be calculated by

$$L_s = L \ [1 + \omega^2 \ CL(10)^{-18}] \tag{154}$$

where L_s is the apparent or observed value of the inductance, C is in micromicrofarads, and L in microhenries. The inductance of a coil is decreased by skin effect, and is increased by capacity. The changes due to these two effects sometimes neutralize each other, and in general, formula (153) gives about as good a value of the high-frequency inductance as can be obtained.

Round Wire.—The low-frequency inductance of a coil wound with round wire can be calculated to much higher precision than that of formula (153) by the use of correction terms. Formula (153) gives strictly, the inductance of the equivalent current sheet, which is a winding in which the wire is replaced by an extremely thin tape, the center of each turn of tape being situated at the center of a turn of wire, the edges of adjacent tapes being separated by an infinitely thin insulation. The inductance of the actual coil is obtained from the current-sheet inductance as follows:

Putting L_s = inductance of equivalent cylindrical current sheet, obtained from (153)

L_0 = inductance of the coil at low frequencies

n = number of turns

a = radius of coil measured out to the center of the wire

D = pitch of winding = distance from center of one wire to the center of the next measured along the axis

b = length of equivalent current sheet = nD

d = diameter of the bare wire

Then $L_0 = L_s - 0.01257 \, na \, (A + B)$ microhenry $\qquad (155)$ in which A is constant, which takes into account the difference in self-inductance of a turn of the wire from that of a turn of the current sheet, and B depends on the difference in mutual inductance of the turns of the coil from that of the turns of the current sheet. The quantities A and B may be interpolated from Tables 11 and 12, page 284, which are taken from Tables 7 and 8 of Bulletin, Bureau of Standards, 8, pages 197–199; 1912.

Example.—A coil of 400 turns of round wire of bare diameter 0.05 cm, wound with a pitch of 10 turns per cm, on a form of such a diameter that the mean radius out to the center of the wire is 10 cm.

$$a = 10, \quad b = nD = 40, \quad n = 400, \quad D = 0.1, \frac{d}{D} = 0.5$$

The value of K corresponding to $\frac{2a}{b} = 0.5$ is 0.8181 (Table 10).

$$L_s = 0.03948 \, (400)^2 \, \frac{100}{40} \, 0.8181 = 0.03948 \times 400\,000 \times 0.8181$$

$$= 12\,919 \text{ microhenries}$$
$$= 0.012919 \text{ henry}$$

$\log 0.03948 = \bar{2}.59638$
$\log 400\,000 = 5.60206$
$\log \ 0.8181 = \bar{1}.91281$

$\overline{\qquad\qquad}$

4.11125

Entering Tables 11 and 12 with $\frac{d}{D} = 0.5$, $n = 400$, we find

$$A = -0.136$$
$$B = \ \ 0.335$$

$\overline{\qquad\qquad}$

$$A + B = \ \ 0.199$$

The correction in (155) is, accordingly

$$0.01257 \, (400) \, 10 \, (0.199) = 9.99 \text{ microhenries.}$$

The total inductance is $12\,919 - 10 = 12\,909$ microhenries.

Example.—A coil of 79 turns of wire of about 0.8 mm bare diameter. The mean diameter is about 22.3 cm and, for determining the pitch, it was found that the distance from the first to the 79th wire was 9.0 cm.

We have, then,

$$a = 11.15, \quad D = \frac{9.0}{78} = 0.115, \quad b = nD = 79 \times 0.115 = 9.12$$

$$\frac{2a}{b} = 2.445, \quad \frac{d}{D} = \frac{0.08}{0.115} = 0.7$$

The value of K is given by Table 10 as 0.4772, so that

$$L_s = 0.03948\,(79)^2\,\frac{(11.15)^2}{9.12}\,0.4772 = 1602.8 \text{ microhenries}$$

$$
\begin{array}{l}
\log 0.03948 = \bar{2}.\,59638 \\
2 \log \quad\quad 79 = 3.\,79526 \\
2 \log \quad 11.15 = 2.\,09454 \\
\log \quad 0.4772 = \bar{1}.\,67870 \\
\hline
\quad\quad\quad\quad 4.\,16488 \\
\log \quad\quad 9.12 = 0.\,95999 \\
\hline
\quad\quad\quad\quad 3.\,20489
\end{array}
$$

For $n = 79$, $\frac{d}{D} = 0.7$, Tables 11 and 12 give

$$A = 0.\,200$$
$$B = 0.\,326$$
$$\overline{\quad\quad\quad}$$
$$(A + B) = 0.\,526$$

The correction is $0.01257 \times 79 \times 11.15 \times 0.526 = 5.8$ microhenries, and the total is 1597.0 microhenries. The measured inductance of this coil is 1595.5.

COIL WOUND WITH WIRE OR STRIP OF RECTANGULAR CROSS SECTION

Approximate values may be obtained for a coil wound with rectangular-section wire or strip by using the simple formula (153), as already explained. More precise values for the low-frequency inductance could be calculated in the same manner as for round wire above, using different values for A and B. It is simpler, however, to use formula (156) below, which applies to the single-layer coil, if the symbols are given the following meaning: $a =$ radius measured from the axis out to the center of the cross section of the wire; $b =$ the pitch of the winding D, multiplied by the number of turns n; $c = w =$ the radial dimension of the wire; $t =$ the axial thickness of the wire. The correction for the cross section of the wire is obtained by using formulas (161) and (162), using $\nu = \frac{w}{D}$, $\tau = \frac{t}{D}$.

Example.—A solenoid of 30 turns is wound with ribbon $\frac{1}{4}$ inch by $\frac{1}{16}$ inch thick, with a winding pitch of $\frac{1}{4}$ inch to form a solenoid of mean diameter 10 inches.

$$\text{Here } a = 5 \times 2.54 = 12.70 \text{ cm}, \quad w = c = \frac{1}{4}(2.54) = 0.635 \text{ cm}$$

$$b = 30 \times \frac{1}{4}(2.54) = 19.05 \text{ cm}, \quad c/b = \frac{1}{30}, \quad D = 0.635$$

$$t = \frac{1}{16}(2.54)$$

for the equivalent coil. Solving this by Rosa's formula (156), using $\frac{2a}{b} = \frac{4}{3}$, $K = 0.6230$ (Table 10), $\frac{b}{c} = 30$, $A_s = 0.693$, $B_s = 0.3218$, we find $L_u = 182.55 \ \mu h$. The value obtained by Stefan's formula (157) is very slightly in error, being 182.5.

To obtain the correction, we have $\nu = \frac{w}{D} = 1$, $\tau = \frac{1}{4}$, and therefore

$$A_1 = \log_e \frac{2}{1.25} = 0.470$$

$$B_1 = -2\left[\frac{29}{30}0.060 + \frac{28}{30}0.018 + \frac{27}{30}0.008 + \frac{26}{30}0.005\right.$$

$$\left. + \cdots + \frac{21}{30}0.001\right] = -0.188$$

so that the correction is $(0.01257) \ 30 \ (12.70) \ (0.285) = 1.36 \ \mu h$, and the total inductance is 183.9.

INDUCTANCE OF POLYGONAL COILS

Such coils, instead of being wound on a cylindrical form, are wrapped around a frame such that each turn of wire incloses an area bounded by a polygon.

No formula has been developed to fit this case, but it is found that the inductance of such a coil (when the number of sides of the polygon is fairly large) may be calculated, within 1 per cent, by assuming that the coil is equivalent to a helix, whose mean radius is equal to the mean of the radii of the circumscribed and inscribed circles of the polygon. That is, if $r =$ the radius of the circumscribed circle, Fig. 182 (which can be measured without difficulty for a polygon for which the number of sides N is an even number), then the modified radius $a_o = r \cos^2 \frac{\pi}{2N}$ is to be used for a in the formulas (153) and (155) of the preceding section.

Examples.—The following table gives the results obtained by this method for some 12-sided polygonal coils, the measured inductance being given for comparison. For $N=12$, $a_o=0.983r$.

Coil	r	a_o	n	D	b	L_o calculated μh	L_o measured μh
A	6.35	6.24	23	0.32	7.3	63.0	61.7
B	8.25	8.10	28	.32	9.0	124.7	126.3
C	11.43	11.22	52	.212	11.0	638.0	630.5
D	11.43	11.22	34	.388	10.8	274.9	274.6
E	13.97	13.73	64	.211	13.1	1119.5	1115.5
F	19.05	18.71	117	.158	18.5	5389	5387

MULTIPLE-LAYER COILS

Different formulas are used for long than for short coils. For long coils of few layers, sometimes called multiple-layer solenoids, the inductance is given, approximately, by

$$L_u = L_s - \frac{0.01257 n^2 ac}{b}(0.693 + B_s) \tag{156}$$

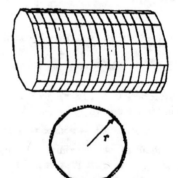

Fig. 182.—*Polygonal coil*

where L_s = inductance, calculated by (153), letting

n = number of turns of the winding

a = radius of coil measured from the axis to the center of cross section of the winding

b = length of coil = distance between centers of turns, times number of turns in one layer

c = radial depth of winding = distance between centers of two adjacent layers times number of layers

B_s = correction given in Table 13, page 284, in terms of the ratio $\frac{b}{c}$

Values obtained by this formula are less accurate as the ratio c/a is greater, and may be a few parts in 1000 in error for values of this ratio as great as 0.25, and $\dfrac{b}{a}$ as great as 5. For accurate results a correction needs to be applied to L_u(see (159) below).

The solution of the problem for short coils is based on that for the ideal case of a circular coil of rectangular cross section. Such a coil would be realized by a winding of wire of rectangular cross

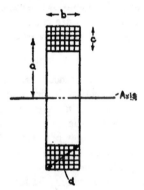

FIG. 183.—*Multiple-layer coil with winding of rectangular cross section*

section, arranged in several layers, with an insulating space of negligible thickness between adjacent wires.

Let $a =$ the mean radius of the winding, measured from the axis to the center of the cross section

$b =$ the axial dimension of the cross section

$c =$ the radial dimension of the cross section

$d = \sqrt{b^2 + c^2} =$ the diagonal of the cross section

$n =$ number of turns of rectangular wire.

Then, if the dimensions b and c are small in comparison with a, the inductance is very accurately given by Stefan's formula, which, for $b > c$, takes the form

$$L_u = 0.01257\,an^2\left[\left(1 + \frac{b^2}{32a^2} + \frac{c^2}{96a^2}\right)\log_e \frac{8a}{d} - y_1 + \frac{b^2}{16a^2}y_2\right]$$

$$= 0.01257\,an^2\left[2.303\left(1 + \frac{b^2}{32a^2} + \frac{c^2}{96a^2}\right)\log_{10} \frac{8a}{d} - y_1 + \frac{b^2}{16a^2}y_2\right] \quad (157)$$

where y_1 and y_2 are constants given in Table 14, page 285.

For disk or pancake coils, $b<c$, and the formula becomes

$$L_u = 0.01257\,an^2\left[\left(1+\frac{b^2}{32a^2}+\frac{c^2}{96a^2}\right)\log_e\frac{8a}{d}-y_1+\frac{c^2}{16a^2}\,y_3\right]$$

$$= 0.01257\,an^2\left[2.303\left(1+\frac{b^2}{32a^2}+\frac{c^2}{96a^2}\right)\log_{10}\frac{8a}{d}-y_1+\frac{c^2}{16a^2}\,y_3\right] \quad (158)$$

in which y_1 and y_3 are given in Table 14, page 285.

The constant y_1 is the same function of both b/c and c/b, so that its argument, in any given case, is the ratio of the smaller dimension to the larger; y_2 and y_3 are functions of c/b and b/c, respectively, the arguments being not greater than unity in either case.

The error due to the neglect of higher order terms in $\frac{b}{a}$ and $\frac{c}{a}$ in formulas (157) and (158) becomes more important the greater the diagonal of the cross section is, in comparison with the mean radius, but even in the most unfavorable case, c/b small, the inaccuracy with values of the diagonal as great as the mean radius does not exceed one-tenth of 1 per cent. The accuracy is greater with disk coils than with long coils, and best of all when the cross section is square.

For long coils (those in which the length b is greater than the mean radius a), the error of formula (157) becomes rapidly greater. In cases where both dimensions of the cross section are large, in comparison with the mean radius, no formulas well adapted to numerical computations are available, but this is not to be regarded as a case of practical importance in radio engineering.

COIL OF ROUND WIRE WOUND IN A CHANNEL OF RECTANGULAR CROSS SECTION

If we suppose that the distance between the centers of adjacent wires in the same layer is D_1, and that the distance between the centers of wires in adjacent layers is D_2, then the dimensions of the cross section of the equivalent coil with uniform distribution of the current over the cross section will be given by $b = n_1 D_1$, $c = n_2 D_2$, where n_1 and n_2 are, respectively, the number of turns per layer, and the number of layers.

The inductance of the equivalent coil calculated by formulas (156), (157), or (158), using these dimensions and the same mean radius as the actual coil, is a very close approximation to the value for the actual coil, unless the percentage of the cross section occupied by insulating space is large.

When such is the case, the correction to the inductance, given in the following formula, may be added:

$$\Delta L = 0.01257 \, an \left[2.30 \log_{10} \frac{D}{d} + 0.138 + E \right] \qquad (159)$$

in which D = distance between centers of adjacent wires

d = diameter of the bare wire

E = a term depending on the number of turns and their arrangement in the cross section. Its value may with sufficient accuracy be taken as equal to 0.017. The correction in (159) should, in any case, be roughly calculated, to see if it need be taken into account.

Example.—Suppose a coil of winding channel $b = c = 1.5$ cm, wound with 15 layers of wire, with 15 turns per layer, the mean radius of the winding being 5 cm. Diameter of bare wire = 0.08 cm.

In this case

$$n = 225, \ d^2 = 4.5, \frac{d^2}{a^2} = \frac{4.5}{25} = 0.18, \ b/c = 1, \ y_1 = 0.8483, \ y_3 = 0.816$$

$$L_u = (0.01257)(5)(225)^2 \left[\left\{ 1 + \frac{3(0.3)^2 + (0.3)^2}{96} \right\} \log . \frac{8}{\sqrt{0.18}} \right.$$

$$\left. - 0.8483 + \frac{(0.3)^2}{16} 0.816 \right]$$

log 8 = 0.90309	2.76310	1.00375 $\log . \frac{8a}{d}$ =	2.9478
$\frac{1}{2}$ log 0.18 = $\bar{1}$.62764	17269	$-y_1 = -$	.8483
	104		
			2.0995
$\log_{10} \frac{8a}{d}$ = 1.27545	2.93683 = log . $\frac{8a}{d} \frac{0.09}{16} 0.816$	=	.0046
			2.104

$\log_{10} 2.104$ = 0.32305

2 $\log_{10} 225$ = 4.70436

$\log_{10} 0.01257$ = $\bar{2}$.09934 L_u = 6694 microhenries.

$\log_{10} 5$ = 0.69897

——————

3.82572

The correction for insulation is found from (159), as follows:

$$\frac{D}{d} = \frac{0.1}{0.08} = \frac{5}{4}, \ \log_{10} \frac{5}{4} = 0.09691, \ \log_e \frac{5}{4} = 0.223$$

$$0.138$$

$$E = 0.017$$

——————

$$0.378$$

correction = (0.01257) (5) (225) 0.378 = 3.34μh

The total inductance is 6697 microhenries = 6.697 millihenries. The correction could, in this case, have been safely neglected.

Example.—A coil of 10 layers of 100 turns per layer, mean radius = 10 cm, the wires being spaced 0.1 cm apart.

For this case $n = 1000$, $a = 10$, $b = 10$, $c = 1$.

Using formula (156) with $\frac{2a}{b} = 2$, $K = 0.5255$, $b/c = 10$

$$L_s = (0.03948) \frac{\overline{1000}^2 \ \overline{10}^2}{10} 0.5255 = 207\ 400 \text{ microhenries.}$$

For the correction, Table 13 gives for $\frac{b}{c} = 10$

$$\begin{array}{c} 0.693 \\ B_s = 0.279 \\ \hline 0.973 \end{array}$$

so that the correction = $(0.01257)10^8 \frac{10}{10} 0.973 = 12\ 200$ and the inductance is

$$L_u = 207\ 400 - 12\ 200 = 195\ 200 \text{ microhenries}$$
$$= 195.2 \text{ millihenries.}$$

The formula (157) gives a value about one part in 900 higher than this.

INDUCTANCE OF A FLAT SPIRAL

Such a spiral may be wound of metal ribbon, or of thicker rectangular wire, or of round wire. In each case, the inductance calculated for the equivalent coil, whose dimensions are measured by the method about to be treated, will generally be as close as 1 per cent to the truth, the value thus computed being too small.

If n wires, Fig. 184, of rectangular cross section are used, whose width in the direction of the axis is w, whose thickness is t, and whose pitch, measured from the center of cross section of one turn to the corresponding point of the next wire is D, then the dimensions of the cross section of the equivalent coil are to be taken as $b = w$, $c = nD$, and as before $d = \sqrt{b^2 + c^2}$.

The mean radius of the equivalent coil is to be taken as $a = a_1 + \frac{1}{2}(n-1)D$, the distance a_1 being one-half of the distance AB (see Fig. 185) measured from the innermost end of the spiral across the center of the spiral to the opposite point of the innermost turn.

The inductance L_u of the equivalent coil is to be calculated using the above dimensions in (158), assuming for n the same number of turns as that of the spiral.

If *round* wire is employed, the same method is used for obtaining the mean radius a and the dimension c, but it is more convenient to take b as zero, and use for the calculation of the inductance of the equivalent coil the special form of (158) which follows when b is placed equal to zero.

$$L_s = 0.01257\ an^2 \left\{ \log_e \frac{8a}{c} - \frac{1}{2} + \frac{c^2}{96a^2} \left(\log_e \frac{8a}{c} + \frac{43}{12} \right) \right\}$$

$$= 0.01257\ n^2 a \left\{ 2.303 \log_{10} \frac{8a}{c} - \frac{1}{2} \right.$$

$$\left. + \frac{c^2}{96a^2} \left(2.303 \log_{10} \frac{8a}{c} + \frac{43}{12} \right) \right\} \qquad (160)$$

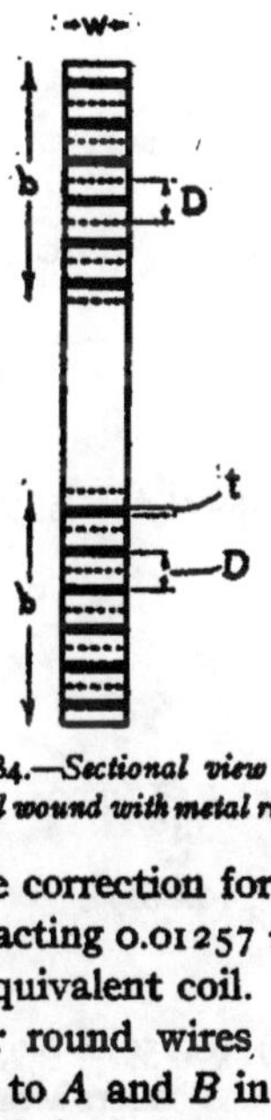

FIG. 184.—*Sectional view of flat spiral wound with metal ribbon*

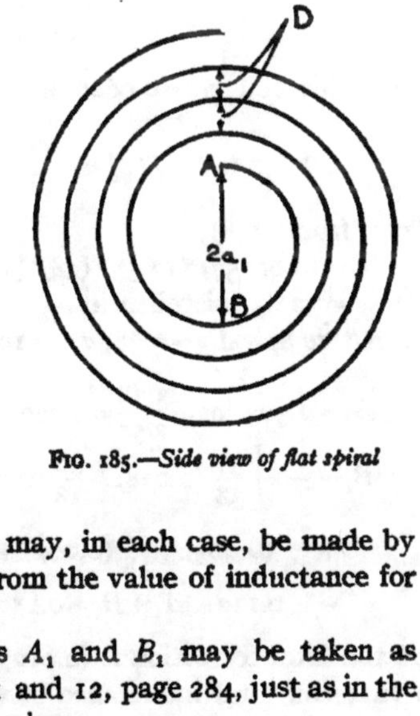

FIG. 185.—*Side view of flat spiral*

The correction for cross section may, in each case, be made by subtracting $0.01257\ na\ (A_1 + B_1)$ from the value of inductance for the equivalent coil.

For round wires the quantities A_1 and B_1 may be taken as equal to A and B in the Tables 11 and 12, page 284, just as in the case of single-layer coils of round wire.

In the case of wire or strip of rectangular cross section the matter is more complicated on account of the two dimensions of the cross section.

If we let $\dfrac{w}{D} = \nu$ and $\dfrac{t}{D} = \tau$, then the quantities involved in the calculation of A_1 and B_1 may be made to depend on these two parameters alone. The equations are then with sufficient accuracy:

$$A_1 = \log_e \frac{\nu + 1}{\nu + \tau} = 2.303 \log_{10} \frac{\nu + 1}{\nu + \tau} \qquad (161)$$

$$B_1 = -2 \left[\frac{n-1}{n} \delta_{12} + \frac{n-2}{n} \delta_{13} + \frac{n-3}{n} \delta_{14} + \ldots + \frac{1}{n} \delta_n \right] \qquad (162)$$

in which δ_{12}, δ_{13}, etc., are to be taken from Table 15, page 285.

Example.—For a spiral of 38 turns, wound with copper ribbon whose cross sectional dimensions are 3/8 by 1/32 inch, the inner diameter was found to be $2a_1 = 10.3$ cm and the measured pitch was found to be 0.40 cm.

The dimensions of the equivalent coil of rectangular cross section are, accordingly,

$$b = 3/8 \text{ inch} = 0.953 \text{ cm,}$$

$$a = \frac{10.3}{2} + \frac{1}{2} \, 37 \, (0.4) = 12.55,$$

$$c = 38 \times 0.40 = 15.2.$$

For this coil $b/c = 0.0627$ which gives (Table 14) $y_1 = 0.5604$,

$y_2 = 0.599$, $\dfrac{d^2}{a^2} = 1.472$, $\log_e \dfrac{8a}{d} = 1.886$.

Hence from (158),

$$L_u = (0.01257) \, (12.55) \, (38)^2 \, [1.015 (1.886) - 0.5604 + 0.055]$$
$$= 323.3 \text{ microhenries.}$$

For this spiral $\nu = 2.38$, $\tau = 0.198$

$$A_1 = 2.303 \log_{10} \frac{3.38}{2.58} = 0.270$$

$$B_1 = -2 \left[\frac{37}{38} \, (0.028) + \frac{36}{38} \, (0.013) + \frac{35}{38} \, (0.007) + \frac{34}{38} \, (0.004) \right.$$

$$\left. + \frac{33}{38} \, (0.003) + \frac{32}{38} \, (0.002) + \frac{31}{38} \, (0.002) + \frac{30}{38} \, (0.001) + \cdots \right]$$

$$= -0.112, \quad A_1 + B_1 = 0.159$$

and the total correction is $(0.01257) \, (38) \, (12.55) \, (0.159) = 0.95 \, \mu h$ so that the total inductance of the spiral is 324.2 microhenries. The measured value was 323.5.

INDUCTANCE OF A SQUARE COIL

Two cases present themselves

(a) A square coil wound in a rectangular cross section.

(b) A square coil wound in a single layer.

MULTIPLE-LAYER SQUARE COIL

Let a be the side of the square measured to the center of the rectangular cross section which has sides b and c, and let n be the total number of turns.

Then

$$L_u = 0.008\ an^2 \left[2.303\ \log_{10} \frac{a}{b+c} + 0.2235 \frac{b+c}{a} + 0.726 \right] \quad (163)$$

If the cross section is a square, $b = c$, this becomes

$$L_u = 0.008\ an^2 \left[2.303\ \log_{10} \frac{a}{b} + 0.447 \frac{b}{a} + 0.033 \right] \quad (164)$$

A correction for the insulating space between the wires may be calculated by equation (159) if we replace $0.01257\ an$ therein by

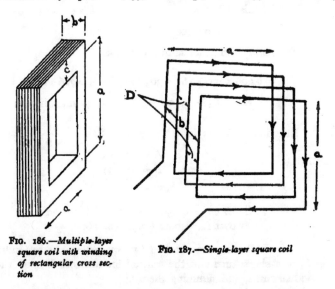

Fio. 186.—*Multiple-layer square coil with winding of rectangular cross section*

Fio. 187.—*Single-layer square coil*

$0.008\ an$. This correction is additive, but will be negligible unless the insulating space between the wires is large.

SINGLE-LAYER SQUARE COIL

Let a = the side of the square, measured to the center of the wire

n = number of turns

D = pitch of the winding, that is, the distance between the center of one wire and the center of the next Fig. 187

$b = nD$

Then

$$L_0 = 0.008 \ an^2 \left[2.303 \ \log_{10} \frac{a}{b} + 0.726 + 0.2231 \ \frac{b}{a} \right]$$
$$- 0.008 \ an \ [A + B] \tag{165}$$

in which A and B are constants having the same meaning as in (155) to be taken from Tables 11 and 12, if the wires are of round cross section. If the wire is a rectangular strip having a dimension t along the axis of the coil and w perpendicular to it, calculate L_u by (163) and correct for cross section by (161) and (162) and Table 15, using 0.008 an $(A_1 + B_1)$.

Example.—Suppose a square coil, 100 cm on a side, wound in a single layer with 4 turns of round wire, 0.1 cm bare diameter, the winding pitch being 0.5 cm.

In this case $n = 4$ $d = 0.1$ $b = 4 \times 0.5 = 2.0$

$$a = 100 \qquad D = 0.5$$

The main term in formula (165) gives

$$0.008 \times 100 \times 16 \left[2.303 \ \log_{10} \frac{100}{2} + 0.726 + 0.004 \right]$$
$$= 128 \ [4.638 + 0.726 + 0.004] = 59.42 \text{ microhenries}$$

Entering Tables 11 and 12, page 284, with $\dfrac{d}{D} = \dfrac{0.1}{0.5} = 0.2$ and $n = 4$,

$$A = -1.053$$
$$B = 0.197$$
$$\text{sum} = -0.856$$

$$0.008 \ an \ [-0.856] = -2.74 \text{ microhenries,}$$
so that $L_u = 59.42 + 2.74 = 62.16$.

This result may be checked by computing the self-inductance L_1 of a single turn and the mutual inductances M_{pq} of the individual turns, and summing them up.
Thus we find

$$4 \ L_1 = 22.65$$
$$6 M_{12} = 21.74$$
$$4 M_{13} = 12.29$$
$$2 M_{14} = \underline{5.50}$$
$$62.18 \text{ microhenries.}$$

Formula (165) applies only when the length b is small compared with the side of the square a.

RECTANGULAR COIL OF RECTANGULAR CROSS SECTION

Let the sides of the rectangle be a and a_1, the dimensions of the cross section b and c, and the number of turns n, $g = \sqrt{a^2 + a_1^2}$

$$L_u = 0.00921 \ (a+a_1) \ n^2 \left[\log_{10} \frac{2aa_1}{b+c} - \frac{a}{a+a_1} \log_{10} (a+g) \right.$$
$$\left. - \frac{a_1}{a+a_1} \log_{10} (a_1+g) \right] + 0.004 \ (a+a_1) \ n^2 \left[2\left(\frac{g}{a+a_1}\right) \right.$$
$$\left. - \frac{1}{2} + 0.447 \frac{(b+c)}{(a+a_1)} \right] \tag{166}$$

Correct for cross section by (159) for round wire.

SINGLE-LAYER RECTANGULAR COIL

Let a and a_1 be the sides of the rectangle, D the pitch of the winding, $b = nD$, and n the number of turns.   Then

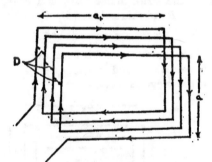

FIG. 188.—*Single-layer rectangular coil.*

$$L_o = 0.00921 \ (a+a_1) \ n^2 \left[\log_{10} \frac{2aa_1}{b} - \frac{a}{a+a_1} \log_{10} (a+g) \right.$$
$$\left. - \frac{a_1}{a+a_1} \log_{10} (a_1+g) \right] + 0.004 \ (a+a_1) \ n^2 \left[\frac{2g}{a+a_1} \right.$$
$$\left. - \frac{1}{2} - 0.447 \frac{b}{a+a_1} \right] - 0.004 \ (a+a_1) \ n \ (A+B) \tag{167}$$

where A and B are to be taken from Tables 11 and 12, if the coil is wound with round wire.   If wound with strip, take $b = nD$ and $c =$ radial thickness of strip.   Calculate L_u by (166) and correct for cross section by (161), (162), and Table 15.

FLAT RECTANGULAR COIL

Let a_0 and a'_0 be the outside dimensions of the coil, measured between centers of the wire, D the pitch of the winding, measured between the centers of adjacent wires (Fig. 189), n the number of complete turns, d the diameter of the bare wire, $c = nD$.

$g = \text{diagonal} = \sqrt{a^2 + a_1^2}$, $a = a_0 - (n-1)D$, $a_1 = a'_0 - (n-1)D$.

Then

$$L_0 = L_u - 0.004 \; n(a + a_1)(A + B)$$

where

$$L_u = 0.009210 \; n^2 \left[(a + a_1) \log_{10} \frac{2aa_1}{c} - a \log_{10}(a + g) \right.$$

$$\left. - a_1 \log_{10}(a_1 + g) \right] + 0.004 \; n^2 \left[2g - \frac{a + a_1}{2} + 0.447 \; c \right] \quad (168)$$

and A and B are constants to be taken from Tables 11 and 12 for round wire. If the coil is wound with rectangular strip, put $b =$ width of the strip, and $c = nD$, and calculate L_u by (166) using for A and B the values A_1 and B_1 of (161) and (162) Table 15.

FLAT SQUARE COIL

If a_0 be here the side of the square, measured between centers of two outside wires, and $a = a_0 - (n-1)D$, the nomenclature being as in the previous section,

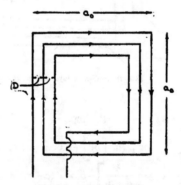

Fig. 189.—*Flat square coil.*

$$L_0 = L_u - 0.008 \; n \; a \; (A + B)$$

in which

$$L_u = 0.008 \; n^2 a \left[2.303 \; \log_{10} \frac{a}{c} + 0.2235 \frac{c}{a} + 0.726 \right] \quad (169)$$

For round wire the constants A and B are given in Tables 11 and 12. If the coil is wound with strip proceed as for rectangular flat coils of strip, above.

Example.—A coil of 4 turns of 0.22 cm stranded wire was found to have $a_0 = 102$ cm, the pitch of the winding being $D = 2.25$ cm. Here

$$a = 102 - 3 \times 2.25 = 95.25$$
$$c = 4 \times 2.25 = 9.0$$

$$L_u = 0.008 \times 16 \times 95.25 \left[2.303 \log_{10} \frac{95.25}{9.0} + 0.2235 \frac{9.0}{95.25} + 0.726 \right]$$

$$= 16 \times 0.762 \, [2.359 + 0.021 + 0.726] = 37.87 \; \mu h$$

For

$$n = 4 \text{ and } \frac{d}{D} = \frac{0.22}{2.25} = 0.098, \text{ Tables 11 and 12 give}$$

$$A = -1.767, \text{ and } B = 0.197$$

the correction is $0.008 \times 4 \times 95.25 \, (-1.570) = -4.79 \, \mu h$ so that $L_0 = 37.87 + 4.79 = 42.66$ microhenries.

The measured value, uncorrected for lead wires was 44.5 microhenries.

DOUBLE FLAT RECTANGULAR COIL

Such a coil consists of two similar flat, rectangular coils, such as are treated in the preceding sections, placed with their axes in the same straight line, and their planes at a distance x apart. The two sections of such a coil may be used either singly, or in series, or in parallel.

The general method of treatment is to obtain the inductance L_1 of the single sections by formula (168) or (166), as described in the preceding sections, and the mutual inductance of the two sections, as shown below.

Then when used in series $L' = 2(L_1 + M)$, and when used in parallel $L'' = \dfrac{L_1 + M}{2}$.

To obtain the mutual inductance, formula (183) or (184) for two equal, parallel rectangles or squares, multiplied by the product of the number of turns of the two, should be used, putting for the dimensions of the rectangles a and a_1 as defined under (168) and (169) and for the distance D in (183) or (184) a modified distance r given by the expression

$$r = kc, \quad c = nD, \quad (x/c \text{ small})$$

in which

$$2.303 \log_{10} k = 2.303 \frac{x^3}{c^3} \log_{10} \frac{x}{c} + \pi \frac{x}{c} - \frac{3}{2} - \frac{3 x^3}{2 c^3} - \frac{1}{12} \frac{x^4}{c^4} \quad (170)$$

When x is not small in comparison with c, r will have to be calculated by the equation

$$\log_{10} r = \frac{x^3}{c^3} \log_{10} x + \frac{1}{2} \left(1 - \frac{x^3}{c^3} \right) \log_{10} (c^3 + x^3) + \frac{\left(2 \frac{x}{c} \tan^{-1} \frac{c}{x} - \frac{3}{2} \right)}{2.303} \quad (171)$$

When the distance x between the planes of the coils is chosen equal to the pitch D of their windings, the calculation of their inductance, when joined in series, may be obtained in a simpler manner. Putting $b = 2D$ and $n_1 = 2n$, the number of turns of the two windings in series,

$$L' = 0.008 \, n_1{}^2 a \left[2.303 \log_{10} \frac{a}{b+c} + 0.2235 \frac{b+c}{a} + 0.726 \right]$$

$$+ 0.008 \, n_1 a \left[2.303 \log_{10} \frac{D}{d} + 0.153 \right] \quad (172)$$

for a square coil, and

$$L' = 0.009210 \, n_1{}^2 \left[(a + a_1) \log_{10} \frac{2 a a_1}{b + c} - a \log_{10} (a + g) \right.$$

$$\left. - a_1 \log_{10} (a_1 + g) \right] + 0.004 \, n_1{}^2 \left[2g - \frac{a + a_1}{2} + 0.447 (b + c) \right]$$

$$+ 0.004 \, n_1 (a + a_1) \left[2.303 \log_{10} \frac{D}{d} + 0.153 \right] \quad (173)$$

for a rectangular coil

$$g = \sqrt{a^2 + a_1{}^2}, \quad d = \text{diameter of bare wire.}$$

Example.—As an example of the use of these formulas, take the case of an actual coil of two sections, each being a flat, square coil of 5 turns of 0.12 cm wire, wound with a pitch of $D = 1.27$ cm, the distance of the planes of the coils being $x = 1.27$ cm. The length of a side of the outside turn was 101 cm.

Putting $n = 5$, $a = 101 - 4 \times 1.27 = 95.9$, $c = 5 \times 1.27 = 6.35$, and $d/D = 0.1$, formula (169) gives $L_1 = 66.28 + 6.14 = 72.42 \mu h$, for a single section.

To obtain the mutual inductance, we find by (170) for

$$\frac{x}{c} = \frac{1.27}{6.35} = 0.2$$

$$2.303 \log_{10} k = 2.303 \times 0.04 (-0.699) + 0.2\,\pi - \frac{3}{2} - \frac{3}{2}(0.04) - \frac{1}{12}(0.0016)$$

$$= -0.0644 + 0.6283 - 1.5 - 0.06 - 0.0001$$

$$= -0.9962$$

$$\log_{10} k = -0.4326 = \bar{1}.5674$$

$$k = 0.3693 \text{ and } r = 0.3693 \times 6.35 = 2.344$$

Putting this value of r in place of D in (184) with $a = 95.9$

$$M = 0.008 \times 5 \times 5 \left[2.303 \times 95.9 \log_{10}\left(\frac{191.8 \times 95.93}{231.5 \times 2.344}\right) + 135.62 \right.$$

$$\left. - 191.86 + 2.34 \right] = 56.82 \ \mu h$$

For the two coils in series, then

$$L' = 2(72.42 + 56.82) = 258.5 \ \mu h$$

and for the parallel arrangement

$$L'' = \frac{72.42 + 56.82}{2} = 64.6 \ \mu h$$

The inductance of the coils in series may also be found by putting $a = 95.9$, $b = 6.35$, $c = 2.54$, $n_1 = 10$ in (163) and (159) and we find $L = 239.8 + 18.8 = 258.6 \ \mu h$ in agreement with the other method.

69. MUTUAL INDUCTANCE

The following formulas for mutual inductance hold strictly only for low frequencies. In general, however, the values will be the same at high frequencies.

TWO PARALLEL WIRES OR BARS SIDE BY SIDE

Let l = length of each wire or bar.

D = distance between centers of the wires.

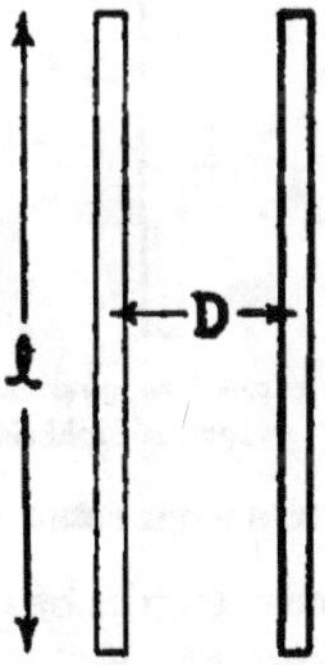

The following expression is exact when the wires have no appreciable cross section, but is sufficiently exact even when the cross section is large if l is

Fig. 190.—*Two parallel wires side by side*

great compared with D. Within these limits the shape is immaterial.

$$M = 0.002l\left[2.303\log_{10}\frac{l+\sqrt{l^2+D^2}}{D}-\sqrt{l^2+D^2}+D\right]\quad(174)$$

$$= 0.002l\left[2.303\log_{10}\frac{2l}{D}-1+\frac{D}{l}\right]\text{nearly.}\quad(175)$$

TWO WIRES END TO END WITH THEIR AXES IN LINE

Let the lengths of the two wires be l and m, their radii being supposed to be small. Then,

$$M = 0.002303\left[l\log_{10}\frac{l+m}{l}+m\log_{10}\frac{l+m}{m}\right]\quad(176)$$

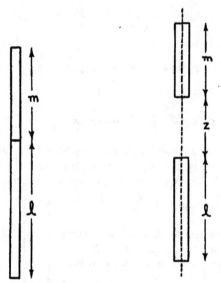

FIG. 191.—*Two wires end to end in same straight line*   FIG. 192.—*Two wires in same straight line but separated*

TWO WIRES WITH THEIR AXES IN THE SAME STRAIGHT LINE BUT SEPARATED

Let their lengths be l and m and the distance between the nearer ends be Z.

$$M = 0.002303\,[(l+m+Z)\log_{10}(l+m+Z)+Z\log_{10}Z$$
$$-(l+Z)\log_{10}(l+Z)-(m+Z)\log_{10}(m+Z)]\quad(177)$$

If AD, AD', AC, AC', etc., represent the distances shown in the figure, the general formula is

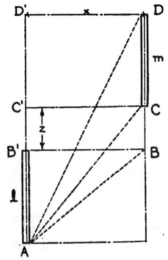

Fig. 193.—*Two wires with axis in parallel lines*

$$M = 0.001151 \left[l \log_{10}\left\{\frac{AD+AD'}{AD-AD'} \times \frac{AC-AC'}{AC+AC'}\right\} \right.$$

$$+ m \log_{10}\left\{\frac{AD+AD'}{AD-AD'} \times \frac{BD-BD'}{BD+BD'}\right\}$$

$$\left. + Z \log_{10}\left\{\frac{AD+AD'}{AD-AD'} \times \frac{AC-AC'}{AC+AC'} \times \frac{BD-BD'}{BD+BD'} \times \frac{BC+BC'}{BC-BC'}\right\}\right]$$

$$- 0.001\ (AD - AC - BD + BC) \qquad (178)$$

the distances being $AD' = l+m+Z$, $AD = \sqrt{x^2+(l+m+Z)^2}$, etc. This formula holds for $Z = 0$, but not when one wire overlaps on the other.

When they overlap, as in Fig. 194,

$$M = M_{1,34} + M_{23} + M_{24} \qquad (179)$$

in which $M_{1,34}$ is to be calculated by the general formula, using $Z = 0$ and putting the segment PV for l and ST for m, while for M_{24} the length VR is put for l and WT for m with $Z = 0$. The mutual inductance M_{33} of the overlapping portions is obtained by (174).

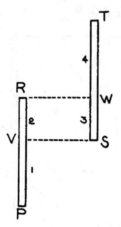

FIG. 194.—*Two wires with axis in parallel
lines; a particular case of Fig. 193*

Special Cases.—For the case shown in Fig. 195

$$M = 0.001 \left[2.303l \log_{10}\left(\frac{l+\sqrt{D^2+l^2}}{D}\right) + D - \sqrt{D^2+l^2} \right] \quad (180)$$

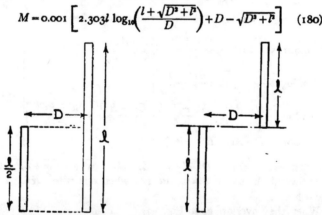

FIG. 195.—*Two wires with axes in parallel*   FIG. 196.—*Two wires with axes in parallel lines,*
lines; another particular case of Fig. 193   *with one end of each on the same perpendicular*

and for the wires of Fig. 196

$$M = 0.001 \left[4.605l \log_{10}\left(\frac{2l+\sqrt{D^2+4l^2}}{l+\sqrt{D^2+l^2}}\right) - \sqrt{D^2+4l^2} \right.$$

$$\left. + 2\sqrt{D^2+l^2} - D \right] \quad (181)$$

MUTUAL INDUCTANCE OF TWO PARALLEL SYMMETRICALLY PLACED WIRES

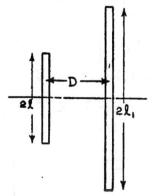

FIG. 197.—*Two parallel symmetrically placed wires*

Putting for the lengths of the two wires $2l$ and $2l_1$, ($2l$ the shorter) and for their distance apart D

$$
\begin{aligned}
M = 0.002 \Bigg[& 2.303(2l)\log_{10}\left\{\frac{l+l_1+\sqrt{(l+l_1)^2+D^2}}{D}\right\} \\
& + 2.303(l_1+l)\log_{10}\left\{\frac{l+l_1+\sqrt{(l+l_1)^2+D^2}}{l_1-l+\sqrt{(l_1-l)^2+D^2}}\right\} \\
& + \sqrt{(l_1-l)^2+D^2} - \sqrt{(l+l_1)^2+D^2} \Bigg]
\end{aligned}
\tag{182}
$$

TWO EQUAL PARALLEL RECTANGLES

Let a and a_1 be the sides of the rectangles and D the distance between their planes, the centers of the rectangles being in the same line, perpendicular to these planes

$$
\begin{aligned}
M = 0.009210 \Bigg[& a \log_{10}\left\{\frac{a+\sqrt{a^2+D^2}}{a+\sqrt{a^2+a_1^2+D^2}} \times \frac{\sqrt{a_1^2+D^2}}{D}\right\} \\
& + a_1 \log_{10}\left\{\frac{a_1+\sqrt{a_1^2+D^2}}{a_1+\sqrt{a^2+a_1^2+D^2}} \times \frac{\sqrt{a^2+D^2}}{D}\right\} \Bigg] \\
& + 0.008\left[\sqrt{a^2+a_1^2+D^2} - \sqrt{a^2+D^2} - \sqrt{a_1^2+D^2}+D\right]
\end{aligned}
\tag{183}
$$

TWO EQUAL PARALLEL SQUARES

If a is the side of each square and D is the distance between their planes, then the preceding formula becomes

$$M = 0.01842 \left[a \log_{10}\left\{ \frac{a + \sqrt{a^2 + D^2}}{a + \sqrt{2a^2 + D^2}} \times \frac{\sqrt{a^2 + D^2}}{D} \right\} \right]$$
$$+ 0.008\left[\sqrt{2a^2 + D^2} - 2\sqrt{a^2 + D^2} + D \right] \quad (184)$$

MUTUAL INDUCTANCE OF TWO RECTANGLES IN THE SAME PLANE WITH THEIR SIDES PARALLEL

$$M = 2(M_{16} + M_{38} + M_{45} + M_{27}) - 2(M_{18} + M_{36} + M_{36} + M_{47}) \quad (185)$$

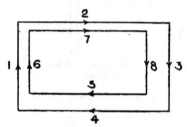

FIG. 198.—*Two rectangles in the same plane with their sides parallel*

the separate mutual inductances being calculated by formula (182), if the sides are symmetrically placed, and by (182) and (178) if that is not the case.

If the rectangles have a common center $M_{16} = M_{38}$, $M_{45} = M_{27}$, $M_{18} = M_{36}$, $M_{36} = M_{47}$ and for the case of concentric squares, we have

$$M = 4(M_{16} - M_{18}) \quad (186)$$

TWO PARALLEL COAXIAL CIRCLES

This is an important case because of its applicability in calculating the mutual inductances of coils (see below).

Let a = the smaller radius (Fig. 199).

A = the larger radius.

D = the distance between the planes of the circles.

Then

$$\frac{r_2}{r_1} = \sqrt{\frac{\left(1 - \dfrac{a}{A}\right)^2 + \dfrac{D^2}{A^2}}{\left(1 + \dfrac{a}{A}\right)^2 + \dfrac{D^2}{A^2}}}$$

must be calculated, and

$$M = F\sqrt{Aa} \qquad (187)$$

where F may be obtained by interpolation in Table 16 for the calculated value of $\frac{r_2}{r_1}$.

$r_1 =$ the longest distance between the circumferences.

$r_2 =$ the shortest distance between the circumferences.

TWO COAXIAL CIRCULAR COILS OF RECTANGULAR CROSS SECTION

If the coil windings are of square, or nearly square, cross section, a first approximation to the mutual inductance is

$$M = n_1 n_2 M_0 \qquad (188)$$

where n_1 and n_2 are the number of turns on the two coils and M_0 is the mutual inductance of two coaxial circles, one located at the center of the cross section of one of the coils and the other at the center of the cross section of the other.

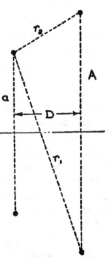

Fig. 199.—*Cross sections of two parallel coaxial circles*

Thus, if

$a =$ mean radius of one coil, measured from the axis to the center of cross section,

$A =$ mean radius, similarly measured, of the other coil,

$D =$ distance between the planes passed through the centers of cross section of the coils, perpendicular to their common axis (Fig. 200).

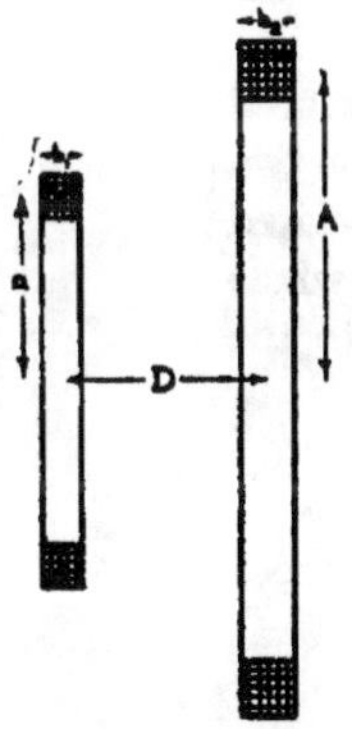

the value M_0 will be computed by formula (187) and Table 16, using the values of a, A, and D, just defined.

If the cross sections of the windings are square, this value will not be more than a few parts in a thousand in error, even with relatively large cross sectional dimensions, except when the coils are close together.

Fig. 200.—*Two parallel coaxial coils with windings of rectangular cross sections*

A more accurate value for coils of square cross section may be obtained by supposing the two parallel circles to remain at the distance D, but to have radii

$$a_1 = a\left(1 + \frac{b_1^2}{24\,a^2}\right) \text{ and } A_1 = A\left(1 + \frac{b_2^2}{24\,A^2}\right) \tag{189}$$

where b_1 and b_2 are the dimensions of the square cross sections corresponding to the coils of mean radius a and A, respectively.

When the correction factors in (189) are only a few parts in 1000, the values of r_2/r_1, and hence F, are very little affected, and the fractional correction to the mutual inductance, to allow for the cross sections, is approximately equal to the geometric mean of the fractional corrections to a and A, so that an estimate of the magnitude of the correction to the mutual inductance may be gained with little labor.

With rectangular cross sections the error from the assumption that the coils may be replaced by equivalent filaments at the center of the cross section is more important than in the case of coils of square cross section and rapidly increases as the axial dimension of one or both of the cross sections is increased, in relation to the distance D between the median planes. The error may, easily, be as great as 1 per cent or more in practical cases.

An estimate of the magnitude of the error, in any case, may be made by dividing the coils up into two or more sections of, as nearly as possible, square cross section, and assuming that each portion of the coil may be replaced by a circular filament at the center of its cross section.

Suppose that coil A is divided into two equal parts, and replaced by two filaments 1, 2, while coil B is likewise replaced by two filaments 3, 4, then, assuming that each filament is associated with a number of turns which is the same fraction of the whole number of turns in the coil as the area of the section is to the whole cross sectional area (one-half in this case) we have

$$M = \frac{n_1}{2}\frac{n_2}{2}\,M_{13} + \frac{n_1 n_2}{4}\,M_{14} + \frac{n_1 n_2}{4}\,M_{23} + \frac{n_1 n_2}{4}\,M_{24}$$
$$= n_1 n_2\left(\frac{M_{13} + M_{14} + M_{23} + M_{24}}{4}\right) \tag{190}$$

in which M_{13} is the mutual inductance of the two circular filaments 1 and 3, etc.

For a discussion of more accurate methods for correcting for the cross section of coils, the reader is referred to Bulletin, Bureau of Standards, **8**, pages 33–43; 1912.

If the coils are of the nature of solenoids of few layers, it is best to use the formulas for the mutual inductance of coaxial solenoids given in the next section.

Example.—Suppose two coils of square cross section 2 cm on a side, the radii being, $a = 20$, $A = 25$, and the distance between their median planes being $D = 10$ cm (Fig. 201). Further, suppose that one coil has 100 turns and the other 500.

Then

$$\frac{r_2}{r_1} = \sqrt{\frac{\left(1 - \frac{20}{25}\right)^2 + \left(\frac{10}{25}\right)^2}{\left(1 + \frac{20}{25}\right)^2 + \left(\frac{10}{25}\right)^2}} = \sqrt{\frac{0.20}{3.40}} = 0.24253$$

From Table 16 we find, corresponding to this value of $\frac{r_2}{r_1}$,

$F = 0.01113$. Therefore, from (187)

$$M_0 = 0.01113\sqrt{25 \times 20} = 0.2489\mu h$$

and

$$M = n_1 n_2 M_0 = 100 \times 500 \times 0.2489$$
$$= 12\ 445 \text{ microhenries}$$
$$= 0.012445 \text{ henry.}$$

FIG. 201.—*Example of two parallel coaxial coils with windings of rectangular cross section*

If we take account of the cross sections we have from (189)

$$a_1 = 20\left(1 + \frac{2^2}{24 \times 20^2}\right) = 20\ (1.00042)$$

$$A_1 = 25\left(1 + \left(\frac{2}{25}\right)^2 \frac{1}{24}\right) = 25\ (1.00027)$$

so that the correction factor to the mutual inductance will be of the order of about $\sqrt{1.00042 \times 1.00027}$, or the mutual inductance should be increased by about 3.5 parts in 10 000 only.

Example.—Fig. 202 shows two coils of rectangular cross section. For coil P, $a = 20$, $b_1 = 2$, $c_1 = 3$, $n_1 = 600$. For coil Q, $A = 25$, $b_2 = 4$, $c_2 = 1$, $n_2 = 400$ and $D = 10$. If, first, we replace each coil by a circular filament at the center of its cross section, we have the same value of M_o as in the previous example, and

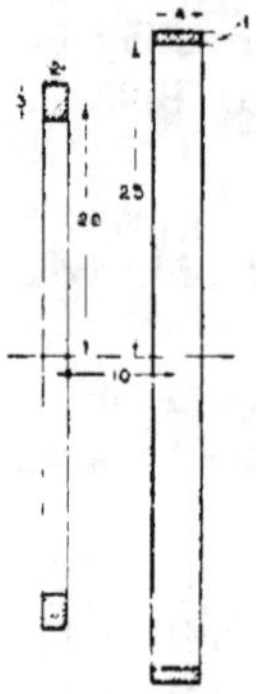

$$M = 600 \times 400 \times 0.2489 \text{ microhenries.}$$

More precise formulas, involving a good deal of computation, show that the true value is

$$M = 600 \times 400 \times 0.249844,$$

so that the approximate value is about 3.8 parts in 1000 too small.

Each coil is then subdivided into two sections and filaments p, q, r, s, imagined to pass through the center of cross section of each of these subdivisions. The data for these filaments are as follows:

FIG. 202.—*Another example of Fig. 200*

Radius	Filaments	a	A	D	r_2/r_1	F
p 19.25	pr	19.25	25	9	0.2365	0.01140
q 20.75	ps	19.25	25	11	.2722	.009872
r 25	qr	20.75	25	9	.2135	.01255
s 25	qs	20.75	25	11	.2506	.01077

We find then

$$M = 600 \times 400 \left\{ \frac{0.2501 + 0.2166 + 0.2858 + 0.2452}{4} \right\} = 600 \times 400 \times 0.24942$$

a result which is 1.7 in 1000 too small.

The increase in accuracy is hardly commensurate with the increased labor.

MUTUAL INDUCTANCE OF COAXIAL SOLENOIDS NOT CONCENTRIC

Gray's formula, given for this case, supposes that each coil approximates the condition of a continuous thin winding, that is, a current sheet.

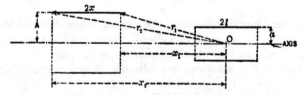

FIG. 203.—*Coaxial solenoids not concentric*

Let a = the smaller radius, measured from the axis of the coil to the center of the wire

A = the larger radius, measured in the same way

$2l$ = length of the coil of smaller radius = number of turns times the pitch of winding

$2x$ = length of the coil of larger radius, measured in the same way

n_1 and n_2 = total number of turns on the two coils

D = axial distance between centers of the coils

$$x_1 = D - x \qquad r_1 = \sqrt{x_1^2 + A^2}$$
$$x_2 = D + x \qquad r_2 = \sqrt{x_2^2 + A^2}$$

Then

$$M = 0.009870 \frac{a^2 A^2 n_1 n_2}{2x \cdot 2l} \left[K_1 k_1 + K_3 k_3 + K_5 k_5 \right] \qquad (191)$$

in which

$$K_1 = \frac{2}{A^2} \left(\frac{x_2}{r_2} - \frac{x_1}{r_1} \right), \ k_1 = 2l$$

$$K_3 = \frac{1}{2} \left(\frac{x_1}{r_1^5} - \frac{x_2}{r_2^5} \right), \ k_3 = a^2 l \left(3 - 4\frac{l^2}{a^2} \right)$$

$$K_5 = -\frac{A^2}{8} \left[\frac{x_1}{r_1^9} \left(3 - 4\frac{x_1^2}{A^2} \right) - \frac{x_2}{r_2^9} \left(3 - 4\frac{x_2^2}{A^2} \right) \right]$$

$$k_5 = a^4 l \left(\frac{5}{2} - 10\frac{l^2}{a^2} + 4\frac{l^4}{a^4} \right)$$

This formula is most accurate for short coils with relatively great distance between them. In the case of long coils it is sometimes necessary to subdivide the coil into two or more parts. The mutual inductance of each of these parts on the other coil having been found, the total mutual inductance is obtained by adding these values.

Example.—

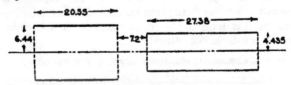

Fig. 204.—*Example of coaxial solenoids not concentric*

$$2x = 20.55 \qquad A = 6.44 \qquad n_1 = 15$$
$$2l = 27.38 \qquad a = 4.435 \qquad n_2 = 75$$

Distance between the adjacent ends of the two solenoids = 7.2 cm.

Then

$$x_1 = 20.89 \qquad\qquad k_1K_1 = 0.04294$$
$$x_2 = 41.44 \qquad\qquad k_2K_2 = .01827$$
$$\qquad\qquad\qquad\qquad k_3K_3 = .00519$$

$$\overline{\qquad\qquad 0.06640}$$

and $M = 0.009870\left(\dfrac{a^2A^2n_1n_2}{2x\ \ 2l}\right)0.06640 = 1.069$ microhenries

$\log 0.009870 = \bar{3}.99432$		
$2 \log a \qquad = 1.29378$	$\log 2x = 1.31281$
$2 \log A \qquad = 1.61778$	$\log 2l = 1.43743$
$\log n_1 n_2 \qquad = 3.05115$	$\overline{\qquad\qquad}$
$\log 0.06640 = \bar{2}.82217$	2.75024

$$2.77920$$
$$2.75024$$

$$\overline{\qquad\qquad}$$

$$0.02896 = \log M$$

Dividing the longer coil into two sections C and D of 37 and 38 turns, respectively, and repeating the calculation for the mutual inductance of these sections on the other coil R (Fig. 204),

For M_{RC}

$$k_1K_1 = 0.04889$$
$$k_2K_2 = .00652$$
$$k_3K_3 = .00005$$

$$\overline{\qquad\qquad}$$

$$0.05546$$

For M_{RD}

$$k_1K_1 = 0.01155$$
$$k_2K_2 = .00061$$

$$\overline{\qquad\qquad}$$

$$0.01216$$

and $M = M_{RC} + M_{RD} = 0.8917 + 0.1956 = 1.087 \ \mu h$.

Further subdivision showed that this last value is not in error by more than 5 parts in 10 000.

The criterion as to the necessity of subdivision is the rapidity with which the terms k_1K_1, k_2K_2, etc., fall off in value. In the first case k_7K_7 and k_8K_8 are not negligible. The expressions for these quantities are not here given because they are laborious to calculate, and it is easier to obtain the value of the mutual inductance by the subdivision method.

COAXIAL, CONCENTRIC SOLENOIDS (OUTER COIL THE LONGER)

The formula here given holds, strictly, only for current sheets. The lengths of the coils should be taken as equal to the number of turns times the pitch of the winding in each case. Then the mutual inductance of the current sheets is not appreciably different from that of the coils.

Let a = smaller radius

A = larger radius

$2x$ = equivalent length of outer coil

$2l$ = equivalent length of inner coil

$g = \sqrt{x^2 + A^2}$ = diagonal.

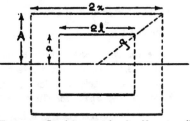

Fig. 205.—*Coaxial concentric solenoids, outer coil begin longer*

Then

$$M = \frac{0.01974 \; a^2 n_1 n_2}{g}\left[1 + \frac{A^2 a^2}{8 g^4}\left(3 - 4\frac{l^2}{a^2}\right)\right] \qquad (192)$$

This formula is more accurate, the shorter the coils and the greater the difference of their radii, but in most practical cases the accuracy is ample. In many cases the second term in (192) is negligible, and it is a good plan to make a preliminary rough calculation of this term to see whether it will need to be considered. In the case of long coils, and of coils of nearly equal radii, the terms neglected in this formula may be as great as 1 per cent. A criterion of rapid convergence is, in general, the smallness of $\frac{a^2 A^2}{g^4}$, but the magnitude of the coefficient $\left(3 - 4\frac{l^2}{a^2}\right)$ and the corresponding coefficients of terms neglected in (192) may in some cases modify this condition for rapid convergence materially.

Example.—

$$2x = 30 \qquad 2l = 5 \qquad g = \sqrt{250} \quad \frac{a^2 A^2}{g^4} = \frac{4}{625}$$

$$A = 5 \qquad a = 4$$

$$n_1 = 300 \qquad n_2 = 200$$

$$0.01974 \; \frac{a^2 n_1 n_2}{g} = 1198.5$$

$M = 1198.5 \; (1 + .00115) = 1199.9$ microhenries.

For the case, however, where

$$2x = 30 \qquad a = 2 \qquad n_1 = 300$$

$$2l = 24 \qquad A = 5 \qquad n_2 = 960$$

although the value of $\frac{a^2A^2}{g^4} = \frac{1}{5000}$ only, the coefficient $\left(3-4\frac{l^2}{a^2}\right)$ $= 141$, (the length of the coil is great compared with its radius) so that the term in $\frac{a^2A^2}{g^4}$ is -0.0282, and investigation of the complete formula shows that the succeeding terms are -0.0127 and -0.0048, so that their neglect will give an error of over 1.5 per cent. (For precision calculations see Bull., Bureau of Standards, 8, pp. 61–64, 1912, for the complete formula.)

CONCENTRIC COAXIAL SOLENOIDS (OUTER COIL THE SHORTER)

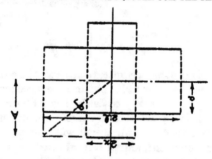

FIG. 206.—*Coaxial concentric solenoids, outer coil bing shorter*

In this case we have to put $g = \sqrt{l^2 + A^2}$, and the formula is

$$M = 0.01974 \frac{a^2n_1n_2}{g}\left[1 + \frac{A^2a^2}{8g^4}\left(3 - 4\frac{x^2}{a^2}\right)\right] \tag{193}$$

which is rapidly convergent in most cases.

70. TABLES FOR INDUCTANCE CALCULATIONS

TABLE 8.—Values of δ in Formulas (132), (134), (137), (138), (140), (141), (142), and (147), for Calculating Inductance of Straight Wires at Any Frequency

z	δ	z	δ
0	0.250	12.0	0.059
0.5	.250	14.0	.050
1.0	.249	16.0	.044
1.5	.247	18.0	.039
2.0	.240	20.0	.035
2.5	0.228	25.0	0.028
3.0	.211	30.0	.024
3.5	.191	40.0	.0175
4.0	.1715	50.0	.014
4.5	.154	60.0	.012
5.0	0.139	70.0	0.010
6.0	.116	80.0	.009
7.0	.100	90.0	.008
8.0	.088	100.0	.007
9.0	.078	∞	.000
10.0	.070		

TABLE 9.—Constants *P* and *Q* in Formulas (141), (142), (144), and (145)

$\frac{2h}{T}$	P	$\frac{l}{2h}$	Q	$\frac{2h}{T}$	P	$\frac{l}{2h}$	Q
0	0	0	1.0000	0.6	0.5136	0.6	1.2918
0.1	0.0975	0.1	1.0499	.7	.5840	.7	1.3373
.2	.1900	.2	1.0997	.8	.6507	.8	1.3819
.3	.2778	.3	1.1489	.9	.7139	.9	1.4251
.4	.3608	.4	1.1975	1.0	.7740	1.0	1.4672
.5	.4393	.5	1.2452				

TABLE 10.—Values of *K* for Use in Formula (153)

$\frac{Diameter}{Length}$	K	Difference	$\frac{Diameter}{Length}$	K	Difference	$\frac{Diameter}{Length}$	K	Difference
0.00	1.0000	−0.0209	2.00	0.5255	−0.0118	7.00	0.2584	−0.0047
.05	.9791	203	2.10	.5137	112	7.20	.2537	45
.10	.9588	197	2.20	.5025	107	7.40	.2491	43
.15	.9391	190	2.30	.4918	102	7.60	.2448	42
.20	.9201	185	2.40	.4816	97	7.80	.2406	40
0.25	0.9016	−0.0178	2.50	0.4719	−0.0093	8.00	0.2366	−0.0094
.30	.8838	173	2.60	.4626	89	8.50	.2272	86
.35	.8665	167	2.70	.4537	85	9.00	.2185	79
.40	.8499	162	2.80	.4452	82	9.50	.2106	73
.45	.8337	156	2.90	.4370	78	10.00	.2033	....
0.50	0.8181	−0.0150	3.00	0.4292	−0.0075	10.0	0.2033	−0.0133
.55	.8031	146	3.10	.4217	72	11.0	.1903	113
.60	.7885	140	3.20	.4145	70	12.0	.1790	98
.65	.7745	136	3.30	.4075	67	13.0	.1692	87
.70	.7609	131	3.40	.4008	64	14.0	.1605	78
0.75	0.7478	−0.0127	3.50	0.3944	−0.0062	15.0	0.1527	−0.0070
.80	.7351	123	3.60	.3882	60	16.0	.1457	63
.85	.7228	118	3.70	.3822	58	17.0	.1394	58
.90	.7110	115	3.80	.3764	56	18.0	.1336	52
.95	.6995	111	3.90	.3708	54	19.0	.1284	48
1.00	0.6884	−0.0107	4.00	0.3654	−0.0052	20.0	0.1236	−0.0085
1.05	.6777	104	4.10	.3602	51	22.0	.1151	73
1.10	.6673	100	4.20	.3551	49	24.0	.1078	63
1.15	.6573	98	4.30	.3502	47	26.0	.1015	56
1.20	.6475	94	4.40	.3455	46	28.0	.0959	49
1.25	0.6381	−0.0091	4.50	0.3409	−0.0045	30.0	0.0910	−0.0102
1.30	.6290	89	4.60	.3364	43	35.0	.0808	80
1.35	.6201	86	4.70	.3321	42	40.0	.0728	64
1.40	.6115	84	4.80	.3279	41	45.0	.0664	53
1.45	.6031	81	4.90	.3238	40	50.0	.0611	43
1.50	0.5950	−0.0079	5.00	0.3198	−0.0076	60.0	0.0528	−0.0061
1.55	.5871	76	5.20	.3122	72	70.0	.0467	48
1.60	.5795	74	5.40	.3050	69	80.0	.0419	38
1.65	.5721	72	5.60	.2981	65	90.0	.0381	31
1.70	.5649	70	5.80	.2916	62	100.0	.0350	....
1.75	0.5579	−0.0068	6.00	0.2854	−0.0059			
1.80	.5511	67	6.20	.2795	56			
1.85	.5444	65	6.40	.2739	54			
1.90	.5379	63	6.60	.2685	52			
1.95	.5316	61	6.80	2633	49			

TABLE 11.—Values of Correction Term A in Formulas (155), (165), (168), and (169)

$\frac{d}{D}$	A	Difference	$\frac{d}{D}$	A	Difference	$\frac{d}{D}$	A	Difference
1.00	0.557	−0.051	0.40	−0.359	−0.052	0.15	−1.340	−0.069
0.95	.506	54	.38	.411	54	.14	1.409	74
.90	.452	57	.36	.465	57	.13	1.483	80
.85	.394	61	.34	.522	61	.12	1.563	87
.80	.334	65	.32	.583	64	.11	1.650	96
0.75	0.269	−0.069	0.30	−0.647	−0.069	0.10	−1.746	−0.105
.70	.200	74	.28	.716	74	.09	1.851	118
.65	.126	80	.26	.790	80	.08	1.969	133
.60	.046	87	.24	.870	87	.07	2.102	154
.55	−.041	95	.22	.957	96	.06	2.256	173
0.50	−0.136	−0.041	0.20	−1.053	−0.051	0.05	−2.439	−0.223
.48	.177	43	.19	1.104	54	.04	2.662	288
.46	.220	44	.18	1.158	57	.03	2.950	405
.44	.264	47	.17	1.215	61	.02	3.355	693
.42	.311	48	.16	1.276	64	.01	4.048	....

TABLE 12.—Values of Correction B in Formulas (155), (165), and (169)

Number of turns, n	B	Number of turns, n	B
1	0.000	40	0.315
2	.114	45	.317
3	.166	50	.319
4	.197	60	.322
5	.218	70	.324
6	0.233	80	0.325
7	.244	90	.327
8	.253	100	.328
9	.260	150	.331
10	.266	200	.333
15	0.286	300	0.334
20	.297	400	.335
25	.304	500	.336
30	.308	700	.336
35	.312	1000	.336

TABLE 13.—Values of B_s for Use in Formula (156)

$\frac{b}{c}$	B_s	$\frac{b}{c}$	B_s
1	0.0000	16	0.3017
2	.1202	17	.3041
3	.1753	18	.3062
4	.2076	19	.3082
5	.2392	20	.3099
6	0.2446	21	0.3116
7	.2563	22	.3131
8	.2656	23	.3145
9	.2730	24	.3157
10	.2792	25	.3169
11	0.2844	26	0.3180
12	.2888	27	.3190
13	.2927	28	.3200
14	.2961	29	.3209
15	.2991	30	.3218

TABLE 14.—Constants Used in Formulas (157) and (158)

b/c or c/b	y_1	Difference	c/b	y_2	Difference	b/c	y_3	Difference
0	0.5000	0.0253	0	0.125	0.002	0	0.597	0.002
0.025	.5253	237						
.05	.5490	434	0.05	.127	5	0.05	.599	3
.10	.5924	386	.10	.132	10	.10	.602	6
0.15	0.6310	0.0342	0.15	0.142	0.013	0.15	0.608	0.007
.20	.6652	301	.20	.155	16	.20	.615	9
.25	.6953	266	.25	.171	20	.25	.624	9
.30	.7217	230	.30	.192	23	.30	.633	10
0.35	0.7447	0.0198	0.35	0.215	0.027	0.35	0.643	0.011
.40	.7645	171	.40	.242	31	.40	.654	11
.45	.7816	144	.45	.273	34	.45	.665	12
.50	.7960	121	.50	.307	37	.50	.677	13
0.55	0.8081	0.0101	0.55	0.344	0.040	0.55	0.690	0.012
.60	.8182	83	.60	.384	43	.60	.702	13
.65	.8265	66	.65	.427	47	.65	.715	14
.70	.8331	52	.70	.474	49	.70	.729	13
0.75	0.8383	0.0039	0.75	0.523	0.053	0.75	0.742	0.014
.80	.8422	29	.80	.576	56	.80	.756	15
.85	.8451	19	.85	.632	59	.85	.771	15
.90	.8470	10	.90	.690	62	.90	.786	15
0.95	0.8480	0.0003	0.95	0.752	0.064	0.95	0.801	0.015
1.00	.8483	....	1.00	.816	....	1.00	.816	....

TABLE 15.—Values of Constants in Formula (162)

ν	Values of δ_{12}						ν	Values of δ_{13}			
	$r=0$	0.1	0.3	0.5	0.7	0.9		$r=0$	0.3	0.6	0.9
0	0.114	0.113	0.105	0.092	0.068	0.030	0	0.022	0.020	0.014	0.004
0.5	.090	.089	.088	.070	.049	.020	0.5	.021	.019	.014	.004
1.0	.064	.064	.059	.050	.034	.013	1.0	.019	.018	.013	.004
1.5	.047	.046	.043	.036	.025	.009	2.0	.015	.015	.010	.003
2.0	.035	.035	.032	.027	.018	.007	4.0	.008	.008	.005	.002
3.0	.022	.022	.020	.017	.011	.004	6.0	.005	.005	.004	.001
4.0	.015	.015	.014	.012	.008	.003	10.0	.003	.003	.002	.005
6.0	.008	.008	.008	.006	.004	.002					
8.0	.006	.006	.005	.004	.003	.001					
10.0	.004	.004	.004	.003	.002	.001					

ν	Values of δ_{14}				ν	Values of δ_{15}			
	$r=0$	0.3	0.6	0.9		$r=0$	0.1	0.5	0.9
0	0.009	0.009	0.006	0.002	0	0.005	0.005	0.004	0.001
1	.009	.008	.006	.002	5	.003	.003	.002	.001
3	.007	.006	.004	.001	10	.002	.002	.001	.000
5	.004	.004	.003	.001					
10	.002	.002	.001	.000					

ν	Values of δ_{16}			ν	Values of δ_{17}			ν	Values of δ_{18}		
	$r=0$ and 0.1	0.5	0.9		$r=0$ and 0.1	0.5	0.9		$r=0$ and 0.1	0.5	0.9
0	0.003	0.003	0.001	0	0.002	0.002	0.001	0	0.002	0.001	0.000
5	.002	.002	.000	5	.002	.001	.000	5	.001	.001	.000
10	.001	.001	.000	10	.001	.001	.000	10	.001	.001	.000

NOTE.—The maximum values of all further values of the δ's are 0.001 or less.

TABLE 16.—Values of F in Formula (187) for the Calculation of the Mutual Inductance of Coaxial Circles

r/n	F	Difference	r/n	F	Difference	r/n	F	Difference
0	∞							
0.010	0.05016	−0.00120	0.30	0.008944	−0.000941	0.80	0.0007345	−0.0000604
.011	4897	109	.31	8503	328	.81	6741	579
.012	4787	100	.32	8175	314	.82	6162	555
			.33	7861	302	.83	5607	531
0.013	4687	−0.00093	.34	7559	290	.84	5076	507
.014	4594	87						
.015	4507	81	0.35	0.007269	−0.000280	0.85	0.0004569	−0.0000484
.016	4426	148	.36	6989	270	.86	4085	460
.018	4278	132	.37	6720	260	.87	3625	437
			.38	6460	249	.88	3188	413
0.020	0.04146	−0.00119	.39	6211	241	.89	2775	389
.022	4027	109						
.024	3918	100	0.40	0.005970	−0.000232	0.90	0.0002386	−0.0000365
.026	3818	93	.41	5738	225	.91	2021	341
.028	3725	86	.42	5514	217	.92	1680	316
			.43	5297	210	.93	1364	290
0.030	3639	−0.00081	.44	5087	202	.94	1074	263
.032	3558	76						
.034	3482	71	0.45	0.004885	−0.000195	0.95	0.00008107	−0.00002351
.036	3411	68	.46	4690	189	.96	5756	2046
.038	3343	64	.47	4501	183	.97	3710	1706
			.48	4318	178	.98	2004	1301
0.040	0.03279	−0.00061	.49	4140	171	.99	703	703
.042	3218	58				1.00	0	
.044	3160	55	0.50	0.003969	−0.000166			
.046	3105	53	.51	3803	160	0.950	0.00008107	−0.00000494
.048	3052	51	.52	3643	156	.952	7613	482
			.53	3487	150	.954	7131	470
0.050	0.03001	−0.00226	.54	3337	146	.956	6661	458
.060	2775	191				.958	6202	446
.070	2584	164	0.55	0.003191	−0.000141			
.080	2420	144	.56	3050	137	0.960	0.00005756	−0.00000436
.090	2276	128	.57	2913	133	.962	5320	421
			.58	2780	128	.964	4899	409
0.100	0.02148	−0.00116	.59	2652	125	.966	4490	397
.11	2032	104				.968	4093	383
.12	1928	96	0.60	0.002527	−0.000120			
.13	1832	89	.61	2407	117	0.970	0.00003710	−0.00000370
.14	1743	82	.62	2290	113	.972	3340	356
			.63	2177	109	.974	2984	341
0.15	0.01661	−0.00075	.64	2068	106	.976	2643	327
.16	1586	71				.978	2316	312
.17	1515	66	0.65	0.001962	−0.000103			
.18	1449	62	.66	1859	99	0.980	0.00002004	−0.00000296
.19	1387	59	.67	1760	96	.982	1708	278
			.68	1664	93	.984	1430	262
0.20	0.01328	−0.00055	.69	1571	90	.986	1168	242
.21	1273	52				.988	926	223
.22	1221	50	0.70	0.001481	−0.000087			
.23	1171	47	.71	1394	84	0.990	0.00000703	−0.00000201
.24	1124	45	.72	1310	81	.992	502	177
			.73	1228	78	.994	326	148
0.25	0.010792	−0.000425	.74	1150	76	.996	177	115
.26	10366	408				.998	062	62
.27	0.009958	388	0.75	0.0010740	−0.0000731			
.28	9570	371	.76	1001	704			
.29	9199	355	.77	0930	680			
			.78	862	653			
			.79	797	628			

DESIGN OF INDUCTANCE COILS

71. DESIGN OF SINGLE-LAYER COILS

The problems of design of single-layer coils may be broadly classified as of two kinds.

(1) Where it is required to design a coil which shall have a certain desired inductance with a given length of wire, the choice of dimensions of the winding and kind of wire to be used being unrestricted within rather broad limits. This class of problems of design includes a consideration of the question as to what

shape of coil will give the required inductance with the minimum resistance.

(2) Given a certain winding form or frame, what pitch of winding and number of turns will be necessary, if a certain inductance is to be obtained.

In the following treatment of the problem the inductance of the coil will be assumed as equal to that of the equivalent cylindrical current sheet. This is allowable, since, in general, the correction for the cross section of the wire will not amount to more than 1 per cent of the total inductance, an amount which may be safely neglected in making the design. The formulas to be given may, of course, be used for making a calculation of the inductance of a given coil. Nevertheless, since their practical use is made to depend upon the interpolation of numerical values from a graph, for accurate calculations formulas (153) and (155) should be used.

The inductances of coils of different size, but of identical shape, and the same number of turns, are proportional to the ratio of their linear dimensions. Every formula for the inductance should, accordingly, be capable of expression in terms of some single chosen linear dimension, all the other dimensions occurring in the formula in pairs in the form of ratios.

Two formulas are here developed, the first applicable to the solution of problems of the first class, giving the inductance in terms of the total length of wire l, the second for problems presupposing a winding frame of given dimensions. Both show the dependence of the inductance on the shape of the coil.

Coil of Minimum Resistance.—The fundamental relations of the constants of a coil are

$$l = 2\pi a n \qquad b = nD$$

$$L_s = 4\pi^2 n^2 \frac{a^2}{b} K \text{ cgs units}$$

the constant K being a function of the shape factor $\frac{2a}{b}$, diameter ÷ length (Table 10, p. 283).

The expression for the inductance may be written as

$$L_s = \frac{2\pi a l n}{b} K$$

and n may be eliminated by substituting for it the expression

$$n = \sqrt{\frac{lb}{2\pi a D}} = \nu \sqrt{\frac{l}{D}},$$

obtained by multiplying together the two expressions involving n above. There results, then,

$$L_s = l\sqrt{\pi\,\frac{2a}{b}\,\frac{l}{D}}\cdot K \quad \text{cgs units}$$

or

$$L_s = \frac{l^{\frac{3}{2}}}{\sqrt{D}}\,\frac{K}{1000}\sqrt{\pi\,\frac{2a}{b}} = \frac{l}{\sqrt{D}}\,F \text{ microhenries.} \qquad (194)$$

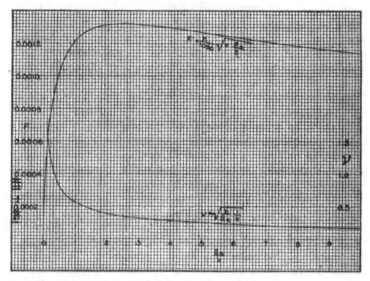

Fig. 207.—(*1*) *Variation of F with different ratios of coil diameter to length; (2) variations of ν with ratios of diameter to length*

To aid in the use of this formula the curve of Fig. 207 has been prepared, which enables the value of $F = \dfrac{K}{1000}\sqrt{\pi\dfrac{2a}{b}}$ to be obtained for any desired value of $\dfrac{2a}{b}$. The formula (194) and the curve enable one to obtain with very little labor the approximate value of the inductance which may be obtained in a coil of given shape with given l and D. On the same figure is also plotted the factor $\nu = \sqrt{\dfrac{b}{\pi\,2a}}$ as a function of $\dfrac{2a}{b}$ (see example below).

Coil Wound on Given Form.—To obtain the second formula, we substitute for n its value $\frac{b}{D}$, and

$$L_s = 4\pi^2 \frac{b^2}{D^2} \frac{a^2}{b} K = 2a\pi \left(\frac{2a}{D}\right)^2 \frac{b}{2a} K \text{ cgs units}$$

or

$$L_s = \frac{(2a)^3}{D^2} \left[\frac{\pi^2}{1000} \frac{b}{2a} K\right] \text{ microhenries} \tag{195}$$

and, finally,

$$\frac{(2a)^3}{L_s D^2} = f \tag{196}$$

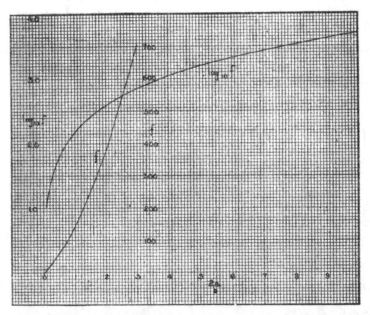

Fig. 208.—*Variation of f and $\log_{10} f$ with $\frac{2a}{b}$*

To aid in making calculations the curves of Fig. 208 have been prepared, which give the values of f and $\log_{10} f = \log_{10}\left[\frac{1000}{\pi^2 K} \frac{2a}{b}\right]$ for different values of $\frac{2a}{b}$. The value of $\log_{10} f$ is plotted, rather than that of f, for large values of $\frac{2a}{b}$, to enable values to be interpolated with greater accuracy.

From formula (194) and Fig. 207 it is at once evident that with a given length of wire, wound with a given pitch, that coil has the greatest inductance, which has such a shape that the ratio $\dfrac{\text{diameter}}{\text{length}} = 2.46$ approximately. Or, to obtain a coil of a certain desired inductance, with a minimum resistance, this relation should be realized. However, although the inductance diminishes rather rapidly for longer coils than this, changes in the direction of making the coil shorter relative to the diameter are not important over rather wide limits. Naturally, other considerations may modify the design appreciably. These other considerations include the distributed capacity of the coil and the variation of resistance with frequency.

Example.—Given the pitch of winding, the shape of the coil $\left(\dfrac{2a}{b}\right)$, and the inductance, to determine the length of wire necessary, the dimensions of the coil and the number of turns.

Assuming $D = 0.2$ cm, $\dfrac{2a}{b} = 2.6$, $L_s = 1000$ microhenries,

By formula (194), $l\tfrac{3}{2} = \dfrac{1000\sqrt{0.2}}{0.001322}$, (the value of $F = 0.001322$ being taken from the curve of Fig. 207) or $l = 4850$ cm. The number of turns may be obtained immediately from the relation $n = \sqrt{\dfrac{l}{D}}\sqrt{\dfrac{b}{2\pi a}} = \nu\sqrt{\dfrac{l}{D}}$ and the graph of ν.

$$\begin{aligned}
\log 1000 &= 3.\\
\tfrac{1}{2}\log 0.2 &= \bar{1}.65052\\
\cline{2-2}
&\quad 2.65052\\
\log F &= \bar{3}.12123\\
\cline{2-2}
{}^{3}/_{2}\log l &= 5.52929\\
\tfrac{1}{2}\log l &= 1.84310\\
\cline{2-2}
\log l &= 3.68619
\end{aligned}$$

Here $n = \sqrt{\dfrac{4850}{0.2}}\ (0.350) = 54.5$ turns, and $b = nD = 10.9$ cm, while $2a = 2.6 \times 10.9 = 28.3$ cm.

If the pitch of the winding had been assumed greater, or a coil of much larger inductance were required, the design of the coil would call for larger dimensions, and cases may arise where the design may prove unsatisfactory, because the coil would be too large. The effect of changing the length and pitch, the shape being taken constant, may be seen from (194), which shows that $L_s \propto \dfrac{l\tfrac{3}{2}}{\sqrt{D}}$, so that a given fractional increase in the length of the wire is more effective in increasing the inductance than the same fractional decrease in the pitch. The number of turns depends on $\sqrt{\dfrac{l}{D}}$ the shape of the coil being kept the same.

Example.—Formula (194) will also enable the question to be answered as to what pitch must be used if a given length of wire is to be wound with a certain shape of coil to give a desired inductance. If the pitch comes out smaller than the diameter of the proposed wire, the assumed length of wire must be increased.

Suppose that an inductance of 10 000 microhenries is desired with 50 meters of wire, the value of $\dfrac{2a}{b}$ being taken as 2.6, as before.

Then

$$\sqrt{D} = \frac{l^{\frac{3}{2}}}{L_s} F = \frac{(5000)^{\frac{3}{2}}\, 0.001322}{10\ 000}, \text{ or } D = 0.00218 \text{ cm,}$$

which is manifestly impracticably small.

The maximum inductance attainable with the given length of wire could be found by solving (194) for L with the smallest practicable pitch substituted for D, that value being used for F, which corresponds to the assumed ratio of diameter to length.

Example.—Suppose we have a winding form of given diameter $2a = 10$ cm, how many turns of wire will have to be used for an inductance of $1000\mu h$ if the winding pitch is taken as 0.2, and what will be the axial length of the winding?

From (196)

$$f = \frac{1000}{1000 \times 0.04} = 25 \text{ or } \log_{10} f = 1.398$$

From Fig. 208 this corresponds to a value of $\dfrac{2a}{b} = 0.225$, or b must be 45 cm, and the number of turns $n = \dfrac{b}{D} = \dfrac{45}{0.2} = 225$. Such a coil would be too long to be convenient. A smaller pitch should be used.

Example.—Suppose we have given the same winding form, and we wish to find what pitch is necessary for an inductance of $1000\mu h$, in order that the length of the coil shall not be greater than the diameter.

For

$$\frac{2a}{b} = 1, f = 148 \text{ (Fig. 208)}$$

and by (196)

$$D^2 = \frac{(2a)^3}{L_s f} = \frac{1000}{1000 \times 148} \text{ or } D = 0.082$$

This is a pretty close winding, showing that the winding form has rather too small a diameter for a coil of this inductance.

Example.—To find the diameter of a winding form to give an inductance of $1000\mu h$, with a shape ratio $\frac{2a}{b} = 2.6$, the pitch being chosen as 0.2 cm.

From (196) we have $(2a)^3 = L_*D^2$.

The value of f for $\frac{2a}{b} = 2.6$ is (from Fig. 208) given by $\log_{10} f = 2.75$ or $f = 565$ approximately. Therefore $(2a)^3 = 1000 \times 0.04 \times 565$, or $2a = 28.2$ cm, which will give $b = 10.85$, $n = 54.2$.

If, instead, the shape is assumed to be given by $\frac{2a}{b} = 1$, then $\log f = 2.17$ or $f = 148$.

$(2a)^3 = 1000 \times 0.04 \times 148$, or $2a = 18.1$ cm $= b$, and $n = 90.5$.

The values of f taken from Fig. 208 are not so precise as could be calculated from the equation (195), but the accuracy should suffice for this kind of work.

72. DESIGN OF MULTIPLE-LAYER COILS

For purposes of design we may neglect the correction for cross section of the wire, formula (159), and operate on formulas (157) and (158) alone.

Two forms of equation have been found useful, the first involving the length of wire in the coil and the second the mean radius of the coil.

Suppose that the length of the winding l, the distance between the centers of adjacent wires D, shape of cross section $\frac{b}{c}$, and the shape ratio of the coil $\frac{c}{a}$, are given. We obtain an expression for n by multiplying together the fundamental equations,

$$n = \frac{bc}{D^2} = \frac{b}{c}\left(\frac{c}{D}\right)^2 \text{ and } n^3 = \frac{l^3}{(2\pi a)^3}$$

which involves ratios of known quantities only.

$$n = \left(\frac{l}{D}\right)^3\left(\frac{c}{a}\right)^3\left(\frac{b}{c}\right)^3\left(\frac{1}{2\pi}\right)^3 \tag{197}$$

In equation (158) the factor $4\pi a n^2 = 2ln$, and if the value of n just found, be introduced, we have finally for $c > b$

$$L = \sqrt[3]{\frac{2}{\pi^2}} \frac{l^{\frac{1}{4}}}{D^{\frac{1}{4}}} \left(\frac{c}{a}\right)^{\frac{1}{4}} \left(\frac{b}{c}\right)^{\frac{1}{4}} \left[\log_e 8 - \log_e \frac{c}{a} - \frac{1}{2} \log_e \left(1 + \frac{b^2}{c^2}\right) - y_1 \right.$$
$$\left. + \frac{c^2}{16a^2} \left\{ y_2 + \frac{1}{6} \left(1 + 3\frac{b^2}{c^2}\right) \left[\log_e \frac{8a}{c} - \frac{1}{2} \log_e \left(1 + \frac{b^2}{c^2}\right) \right] \right\} \right] \quad (198)$$

and for $b > c$

$$L = \sqrt[3]{\frac{2}{\pi^2}} \frac{l^{\frac{1}{4}}}{D^{\frac{1}{4}}} \left(\frac{c}{a}\right)^{\frac{1}{4}} \left(\frac{b}{c}\right)^{\frac{1}{4}} \left[\log_e 8 - \log_e \frac{c}{a} - \log_e \frac{b}{c} - \frac{1}{2} \log_e \left(1 + \frac{c^2}{b^2}\right) - y_1 \right.$$
$$\left. + \frac{c^2}{16a^2} \frac{b^2}{c^2} \left\{ y_2 + \frac{1}{2} \left(1 + \frac{c^2}{3b^2}\right) \left[\log_e \frac{8a}{c} - \log_e \frac{b}{c} - \frac{1}{2} \log_e \left(1 + \frac{c^2}{b^2}\right) \right] \right\} \right] \quad (199)$$

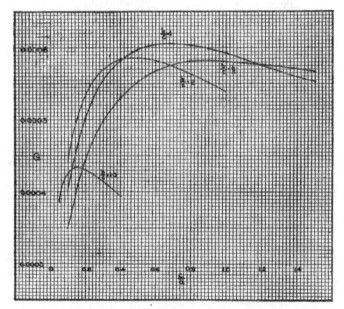

Fig. 209.—*Values of* (G) *for given values of* $\frac{c}{a}$ *and* $\frac{b}{c}$

Both of these equations may be written in the form

$$L = \frac{l^{\frac{1}{4}}}{D^{\frac{1}{4}}} G \text{ microhenries} \quad (200)$$

in which G is a factor whose value for given values of $\frac{c}{a}$ and $\frac{b}{c}$ may be taken from the curves of Fig. 209.

When l is known

$$a = \sqrt[3]{\frac{l}{2\pi} \frac{c}{b} \frac{D^2}{(c/a)^2}}$$ (201)

From these curves one can see that, for a square cross section, $b/c = 1$, the inductance of a given length of wire is a maximum for a value of $\frac{c}{a}$ equal to about $\frac{2}{3}$. Investigation shows that this point is, more exactly, $c/a = 0.662$; that is, for a mean diameter of coil $= 3.02$ times the side of the cross section. Further, for a given resistance and shape of coil, the square cross section gives a greater inductance than any other form.

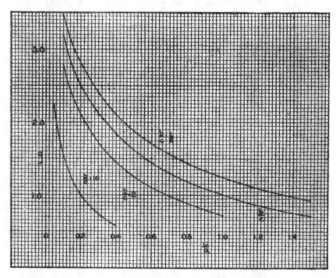

Fig. 210.—*Values of* (g) *for given values of* $\frac{c}{a}$ *and* $\frac{b}{c}$

The second design formula supposes that the dimensions a, c, and $\frac{b}{c}$ of the winding form are given, together with the pitch of the winding. The expressions (157) and (158) for the inductance may then be written

$$L = 0.01257 \, a \, \frac{b^2}{c^2}\left(\frac{c}{D}\right)^4 g \text{ microhenries}$$ (202)

$$= 0.01257 \, an^2 g$$ (203)

The curves of Fig. 210, which give g for different values of $\frac{c}{a}$ and $\frac{b}{c}$ allow of interpolation of the proper value in any given case.

Example.—Suppose we have a wire of such a size that it may be wound 20 turns to the centimeter, and we wish to design a coil to have an inductance of 10 millihenries, to have a square cross section and such a mean radius as to obtain the desired inductance with the smallest resistance (smallest length of the wire).

The latter condition requires that $\frac{c}{a}=0.662$. The given quantities are $D=0.05$ cm, $b/c=1$. From Fig. 209 we find that $G=0.000606$, so that (200) becomes $10\ 000 = \frac{l^4}{(0.05)^3}\ 0.000606$, from which $l=6458$ cm or. 64.58 meters of wire.

$$2/3 \log D = 1.13265 \qquad \text{From the fundamental equation} \quad (201)$$

$$\log \frac{10^7}{0.606} = 7.21753$$

$$\overline{5/3 \log l = 6.35018}$$
$$1/3 \log l = 1.27004$$

$$a = \sqrt[3]{\frac{l}{2\pi} \cdot \frac{c}{b} \cdot \frac{D^2}{(c/a)^3}}$$

$$\overline{2 \log l = 7.62022}$$
$$\log l = 3.81011 \qquad\qquad = 1.80$$

and thence $b=c=0.662 \times 1.80 = 1.19$, and $n=\frac{bc}{D^2}=\frac{(1.19)^2}{0.0025}=570$.

This coil is rather too small to allow of its dimensions being accurately measured.

If wire of double the pitch is used, the design works out with the following results

$$l = 85.22 \text{ meters} \qquad c = b = 2.08$$
$$n = 432 \qquad\qquad a = 3.18$$

which is more suitable.

Example.—We have a form whose dimensions are $2a = 10$, $c = 3$, $b = 2.4$, wound with wire of such a size that there are 10 turns per cm; that is, $D = 0.1$. What is the inductance obtained and what length of wire is used?

$$n = \frac{bc}{D^2} = \frac{3 \times 2.4}{0.01} = 720$$

From Fig. 210 the interpolated value of g for $\frac{b}{c} = 0.8$, $c/a = 0.6$, is 1.54 (calculated directly from (158) = 1.552). Accordingly,

$$L = 0.01257 \times 5 \times \overrightarrow{720}^2 \times 1.54 = 50\ 160\ \mu h.$$
$$= 50.16 \text{ millihenries.}$$

The length of wire is $l = 2\pi\, an = 10\ \pi\ 720 = 22\ 600$ cm
$$= 226 \text{ meters.}$$

Example.—The same formula might be used to answer the question, How many turns would have to be wound (completely filling this cross section) in order to obtain a desired inductance, say 20 millihenries. From (203),

$$n^2 = \frac{L}{0.01257\ ag} = \frac{20\ 000}{(0.01257)\ 5\ (1.54)} = 206\ 500$$

or n would be 454, which would mean that

$$D^2 = \frac{bc}{454} = \frac{7.20}{454} = 0.0158$$

or $D = 0.126$, so that the wire would have to wind about 8 turns to the centimeter.

The skin effect and capacity between the layers of the wire are larger in this kind of coil than in the other forms previously considered. A multiple layer coil is therefore to be regarded as undesirable in radio work, and if it be used the cross section should be made small relative to the mean radius.

73. DESIGN OF FLAT SPIRALS

The design of a flat spiral differs from that of a multiple layer coil in that the actual width b of the tape used (not b/c) is supposed to be a given quantity.

The fundamental equations are

$$n = \frac{c}{D} \text{ and } n = \frac{l}{2\pi a},$$

which, on multiplication, give

$$n = \sqrt{\frac{1}{2\pi} \cdot \frac{c}{a} \cdot \frac{l}{D}} \tag{204}$$

and this introduced into the expression $4\pi a n^2 = 2ln$ gives finally

$$L = \frac{l^{\frac{3}{2}}}{\sqrt{D}}\sqrt{\frac{2}{\pi}\frac{c}{a}}\left[\left\{\log_e 8 - \log_e \frac{c}{a} - \tfrac{1}{2}\log_e \left(1 + \frac{b^2}{c^2}\right) - y_1\right\}\right.$$

$$\left. + \frac{1}{16}\frac{c^2}{a^2}\left[y_3 + \frac{1}{6}\left(1 + \frac{3b^2}{c^2}\right)\left\{\log_e \frac{8a}{c} - \tfrac{1}{2}\log_e \left(1 + \frac{b^2}{c^2}\right)\right\}\right]\right]$$

$$= \frac{l^{\frac{3}{2}}}{\sqrt{D}}H \text{ microhenries.} \tag{205}$$

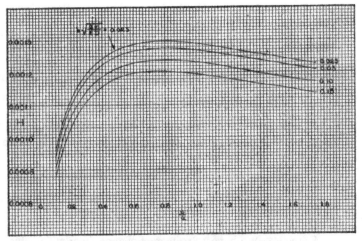

FIG. 211.—*Value of (H) for given values of $\frac{c}{a}$ and $\frac{b}{c}$*

The factor H, which may be determined from the curves of Fig. 211 is a function of c/a and b/c. The latter quantity may be expressed in terms of the known quantities by the equation

$$\frac{b}{c} = b\sqrt{\frac{2\pi}{lD}} \div \sqrt{\frac{c}{a}} \tag{206}$$

Accordingly, the curves are plotted with H as ordinates, c/a as abscissas, and $b\sqrt{\frac{2\pi}{lD}}$ as parameter.

An important deduction which may be made from the curves is that for the maximum inductance with a given length of tape the ratio c/a should be about $\frac{3}{4}$, which means that the opening of the spiral should have a radius nearly as great as the dimension across the turns of the spiral. This point in design is in agreement with the practical observation that turns in the center of the spiral add a disproportionate amount to the high-frequency resistance of the spiral.

Example.—Find the length of tape 0.6 cm wide, wound with a pitch of 0.6 cm, to give an inductance of 200 μh, assuming such proportions that $c/a = 1$. Work out the design.

Since l is not known, the parameter $b \sqrt{\frac{2\pi}{lD}}$ is not known. Assume a value of 0.1 for the latter. Then for the value $c/a = 1$ the curve (Fig. 211) gives $H = 0.00123$.

Thence $l^{\frac{3}{2}} = \frac{200\sqrt{0.6}}{0.00123}$ or $l = 3287$ cm. With this value of l, the parameter is $0.6 \sqrt{\frac{2\pi}{1972}}$, or 0.0339, to which the value $H = 0.00128$ corresponds (with $\frac{c}{a} = 1$). Repeating the calculation of l with this value of H, we find $l = 3370$ cm as a second approximation. The next approximation gives a parameter of 0.0335 and the values of H and l are sensibly unchanged.

Using this parameter in (206), $\frac{b}{c} = 0.0335$ or $c = \frac{0.6}{0.0335} = 17.9$ and the value of $a = 17.9$ likewise. The number of turns will be $n = \frac{17.9}{0.6}$ = about 30.

Example.—We have 17.50 meters of tape 1 cm wide, which we wind with a pitch of 0.5 cm, to such a shape that $c/a = 0.8$.

Here $D = 0.5, l = 1750$ cm, $b = 1$. The parameter is $\sqrt{\frac{2\pi}{875}} = 0.0847$, to which, for $c/a = 0.8$, $H = 0.001248$ corresponds.

$$L = \frac{(1750)^{\frac{3}{2}}}{\sqrt{0.5}} 0.001248 = 129.2 \ \mu h$$

$$\frac{b}{c} = \frac{0.0847}{\sqrt{0.8}} = 0.0947, \text{ by equation (206)}$$

$$c = \frac{1}{0.0947} = 10.56 \text{ cm.}$$

$$a = \frac{10.56}{0.8} = 13.2$$

and the number of turns, $n = \frac{10.56}{0.5} = 21$ nearly.

Example.—The problem may arise as to how closely the tape in the preceding case would have to be wound, still keeping $\frac{c}{a}=0.8$, to obtain an inductance of 200 μh.

Changing the pitch D will change the parameter of the curves, and hence H. The changes in the latter will not be important, for small changes in D, so that to a first approximation the inductance will change inversely as $\sqrt{D}$.

Therefore

$$\sqrt{\frac{D}{0.5}} = \frac{129.2}{200}, \text{ or } D = 0.2086 \text{ cm.}$$

Calculating the parameter with this value we find 0.1312, and thence $H = 0.001216$, so that the second approximation is $\sqrt{D} = \frac{(1750)^{\frac{1}{2}}}{200}(0.001216)$, and $D = 0.1981$, and another approximation is 0.197, the parameter being 0.1346. The dimensions are found from

$$\frac{b}{c} = \frac{0.1346}{\sqrt{0.8}} = 0.1505 \qquad c = \frac{1}{0.1505} = 6.64$$

$$a = \frac{c}{0.8} = 8.30 \qquad\qquad n = \frac{6.649}{0.197} = 34 \text{ nearly.}$$

HIGH-FREQUENCY RESISTANCE

74. RESISTANCE OF SIMPLE CONDUCTORS

Two principal causes act to increase the resistance of a circuit carrying a current of high frequency, above the value of its resistance with direct current, viz, the so-called skin effect and the capacity between the conductors. This section deals exclusively with the skin effect or change of resistance caused by change of current distribution within the conductor. (See sec. 3.)

Unfortunately, formulas for the skin effect are available only for the most simple circuits; and for other very common cases in practice only qualitative indications of the magnitude of the increase in resistance can be given.

In what follows

R = the resistance at frequency f

R_0 = the resistance with direct current or very low frequency alternating current.

The quantity of greatest practical interest is not R, but the resistance ratio $\frac{R}{R_0}$. Given this ratio for the desired frequency and the easily measured direct-current resistance, the high-frequency resistance follows at once.

The skin effect in a conductor always depends, in addition to the thickness of the conductor, on the parameter $\sqrt{\frac{2\mu f}{\rho}}$, in which μ = permeability of the material, f = frequency of the current, ρ = the volume resistivity in microhm-cms, so that as far as skin effect is concerned, a thick wire at low frequencies may show as great a skin effect as a thin one at much higher frequency.

The skin effect is greater in good conductors than in wires of high resistivity, and conductors of magnetic material show an exaggerated increase of resistance with frequency.

Cylindrical Straight Wires.—For this case accurate values of the resistance ratio are given by the formula and tables here given.

If d is the diameter of the cross section of the wire in cm, the quantity

$$x = \pi d \sqrt{\frac{2\mu f}{\rho}} \tag{207}$$

must be calculated (or, in the case of copper, obtained for the desired frequency from Table 19, p. 311 and formula (209)). Knowing the value of x, the value of $\frac{R}{R_0}$ may be taken at once from Table 17, page 309, which gives the value of $\frac{R}{R_0}$ directly for a wide range of values of x.

Table 19 gives values of

$$a_0 = 0.1071 \sqrt{f} \tag{208}$$

for a copper wire at 20° C, 0.1 cm in diameter, and at various frequencies. The value of x for a copper wire of diameter d in cm is

$$x_0 = 10 d a_0 \tag{209}$$

For a material of resistivity ρ and permeability μ, the parameter x may also be simply obtained from the value which holds for a copper wire of the same diameter, by multiplying the latter value by $\sqrt{\mu \frac{\rho_0}{\rho}}$.

The range of Table 19 may be considerably extended by remembering that a is proportional to $\sqrt{f}$ or $\sqrt{\frac{1}{\lambda}}$, where λ is the wave length.

Table 18, page 310, will be found useful, when it is desired to determine what is the largest diameter of wire of a given material, which has a resistance ratio of not more than 1 per cent greater than unity. These values are, of course, based on certain assumed values of resistivity; temperature changes and differences of chemical composition will slightly alter the values. In the case of iron wires μ is the effective permeability over the cycle. This will, in general, be impossible to estimate closely. The values given show plainly how important is the skin effect in iron wires.

For a resistance ratio only one-tenth per cent greater than unity the values in Table 18 should be multiplied by 0.55, and for a 10 per cent increase of the high-frequency resistance the diameters given in the table must be multiplied by 1.78.

The formulas above given apply only to wires which are too far away from others to be affected by the latter. For wires near together, as, for example, in the case of parallel wires forming a return circuit, the mutual effect of one wire on the other always increases the ratio $\frac{R}{R_0}$. No formula for calculating this effect is available, but it is only for wires nearly in contact that it is important. At distances of 10 to 20 cm the mutual effect is entirely negligible.

Tubular Conductors.—The resistance ratio of tubular conductors in which the thickness of the walls of the tube is small in comparison with the mean diameter of the tube, may be calculated by the theoretical formula for an infinite plane of twice the thickness of the walls of the tube.

The value of the resistance ratio for this case may be obtained directly from Table 20, page 311, in terms of the quantity

$$\beta = x\tau\sqrt{2}, \tag{210}$$

where

$\tau =$ the thickness of the walls of the tube in cm
$x =$ the parameter defined in formula (207).

For copper tubes the parameter β_0 may be obtained very simply from the values of a_0 in Table 19, page 311, and the relation $\beta_0 = 10\sqrt{2}\,\tau\,a_0$.

For values of β greater than 4 no table is necessary, since we have simply, with an accuracy always greater than one-tenth of 1 per cent,

$$\frac{R}{R_0} = \beta \qquad (211)$$

Sufficient experimental evidence is not available to indicate an accurate method of procedure in the case of tubing where the ratio of diameter to wall thickness is not large. Measurements with tubing in which this ratio is as small as two or three indicate that approximate values of $\frac{R}{R_0}$ for this case may be calculated by using for τ, in the calculation of the parameter β, a value equal to two-thirds of the actual thickness of the walls of the tube.

Tubing which is very thin in comparison with its radius has, for the same cross section, a smaller high-frequency resistance than any other single conductor. For this reason galvanized-iron pipe is a good form of conductor for some radio work, the current all flowing in the thin layer of zinc. A conductor of smaller resistance than a tube of a certain cross section is obtained by the use of very fine strands separated widely from one another; there are practical difficulties, however, in making the separation great enough.

In a return circuit of tubular conductors the distance between the conductors should be kept as great as 10 or 20 cm. For tubular conductors nearly in contact the resistance ratio may be double that for a spacing of a few centimeters.

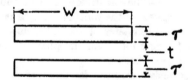

FIG. 212.—*Cross section of strip conductors forming a return circuit with narrow surfaces in the same plane*

Strip Conductors.—If two strips form together a return circuit and they are so placed that there is only a small thickness of dielectric between the wider face of one and the same face of the other (Fig. 212), the resistance ratio may be calculated by formula (210), using for τ the actual thickness of the strip.

As the thickness of the insulating space between the plates is increased, the accuracy of the formula decreases, but the error does not amount to more than a few per cent for values of this thickness as great as several centimeters.

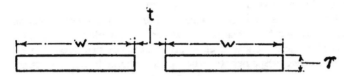

Fɪɢ. 213.—*Cross section of strip conductors forming a return circuit with wide surfaces in the same plane*

For a return circuit of strips placed with their wider faces in the same plane (Fig. 213), no formula is available. This is an unfavorable arrangement. As the distance t is reduced below a few centimeters the ratio $\frac{R}{R_0}$ increases rapidly and with the strips very close together may be as great as twice the value for the arrangement of Fig. 212.

For single strips—that is, for return circuits in which the distance between the conductors is so great that there is no appreciable mutual effect between the conductors—formula (210) is inapplicable owing to "edge effect"—the effect of the magnetic field produced by the current in the center of the strip upon the outer portions of the cross section.

Thus the resistance ratio $\frac{R}{R_0}$ is greater in a wide strip than in a narrow one of the same thickness, and in every case the resistance ratio is greater than for the two juxtaposed strips of Fig. 212. For $\frac{R}{R_0}$ between 1 and 1.5, the increase over formula (210) is usually not greater than 10 per cent.

Strips of square, or nearly square, cross section have values of $\frac{R}{R_0}$ not very different from those which hold for round conductors of the same area of cross section, the values being greater for the square strip than for the round conductor whose diameter is equal to the side of the square.

Simple Circuits of Round or Rectangular Wire.—The ratio of the resistance at high frequencies to that with direct current may be accurately obtained from Table 17, page 309, for circles or rectangles of round wire and in fact for any circuit of which the length is great compared with the thickness of the wire, provided no considerable portions of the circuit are placed close together. In the latter case, the resistance ratio is somewhat increased beyond the value calculated by the previous method and by an amount which can not be calculated.

The resistance ratio for a circuit of wire of rectangular section may be treated by the same method as for a single strip. If portions of the circuit are in close proximity, the precautions mentioned for two strips near together (p. 303) should be borne in mind.

75. RESISTANCE OF COILS

Single-Layer Coil; Wire of Rectangular Cross Section.—The only case for which an exact formula is available is that of a single-layer winding of wire of rectangular cross section with an insulation of negligible thickness between the turns, the length of the winding being assumed to be very great compared with the mean radius, and the latter being assumed very great compared with the thickness of the wire.

If R = the resistance at high frequency

R_0 = the resistance to direct current

τ = the radial thickness of the wire

b = the axial thickness of the wire

ρ = the volume resistivity of the wire in microhm-cm

ρ_0 = the volume resistivity of copper

μ = the permeability of the wire

D = the pitch of the winding,

then $\dfrac{R}{R_0}$ may be obtained directly from Table 20, page 311, having calculated first the quantity $\beta = 10\tau\sqrt{2}\,a$, in which $a = 0.1985\sqrt{\dfrac{\mu f}{\rho}}$. Values of a_0 for copper are given in Table 19, page 311, and the value of a for any other material is obtained from a_0 by the relation $a = a_0\sqrt{\mu\dfrac{\rho_0}{\rho}}$. For values of β greater than are included in Table 20 we have simply $\dfrac{R}{R_0} = \beta$.

In practice the ideal conditions presupposed above will not be realized. To reduce the value calculated for the idealized winding corrections need to be applied: (1) For the spacing of the wire, (2) for the round cross section of the wire, (3) for the curvature of the wire, (4) for the finite length of the coil.

Correction for Pitch of the Winding.—To take into account the fact that the pitch of the winding is not in general equal to the axial breadth of the wire an approximation is obtained if for β the argument

$$\beta' = \beta\sqrt{\frac{b}{D}} \text{ is substituted.}$$

For values of D greater than about $3b$ the values of $\frac{R}{R_0}$ thus obtained are too small.

Correction for the Round Cross Section of the Wire.—For coils of round wire only empirical expressions are known, and more experimental work is desirable.

To obtain an accuracy of perhaps 10 per cent in the resistance ratio the following procedure may be used:

Calculate first by (210) and Table 20, page 311, the resistance ratio $\frac{R'}{R_0'}$, supposing the coil to be wound with wire of square cross section of the same thickness as the actual diameter, taking into account the correction for the pitch of the winding. Then the resistance ratio $\frac{R}{R_0}$ for a winding of round wire will be found by the relation

$$\frac{R}{R_0} = 1 + 0.59\left[\frac{R' - R_0'}{R_0'}\right] \tag{212}$$

Effect of Thickness of the Wire.—Although formula (210) holds only for a coil whose diameter is very great in comparison with the thickness of the wire, the error resulting from non-fulfillment of this condition will, in practical cases, be small compared with the other corrections and may be neglected.

Correction for Finite Length of the Coil.—For short coils the resistance ratio is greater than for long coils of the same wire, pitch, and radius, due to the appreciable strength of the magnetic field close to the wires on the outside of the coil.

No formulas are available for calculating this effect, but experiment seems to show that for short coils of thick wire at radio frequencies the resistance ratio may be expressed by

$$\frac{R}{R_0} = \frac{A}{\sqrt{\lambda}} + \frac{B}{\lambda^3} \tag{213}$$

in which the first term represents the value as calculated by the formulas of the preceding section for long coils, while the constant of the second term has to be obtained by experiment. At long wave lengths the first term will predominate, but at very short wave lengths the second term may be equal or even larger than the first.

For round copper wires we may obtain the constant A by the relation $A = 15\,500\,dR_0$.

Multiple-Layer Coils.—For this case no accurate formulas have been derived. Experiment shows that the resistance ratio is much greater for a multiple-layer coil than for a single-layer coil of the same wire. Furthermore, the capacity of such a coil has, as already pointed out, a large effect on the resistance of the coil. Consequently, it is usually impossible to calculate even an approximate value for the change of resistance with frequency. At very high frequencies losses in the dielectric between the wires may cause an appreciable increase in the effective resistance of the coil. This effect is proportional to f^2.

76. STRANDED-WIRE

The use of conductors consisting of a number of fine wires to reduce the skin effect is common. The resistance ratio for a stranded conductor is, however, always considerably larger than the value calculated by Table 19, page 311, and Table 17, page 309, for a single one of the strands. Only when the strands are at impracticably large distances from one another is this condition even approximately realized.

Formulas have been proposed for calculating the resistance ratio of stranded conductors,[36] but although they enable qualitatively correct conclusions to be drawn as to the effect of changing the frequency and some of the other variables, they do not give numerical values which agree at all closely with experiment. The cause for this lies, probably, to a large extent in the importance of small changes in the arrangement of the strands. The following general statements will serve as a rough guide as to what may be expected for the order of magnitude of the resistance ratio as an aid in design, but when a precise knowledge of the resistance ratio is required in any given case it should be measured. (See methods given in sections 46 to 50.)

Bare Strands in Contact.—The resistance ratio of n strands of bare wire placed parallel and making contact with one another is found by experiment to be the same as for a round solid wire

---

[36] See references 112 to 123 of the Bibliography.

which has the same area of cross section as the sum of the cross-sectional areas of the strands; that is, *n* times the cross section of a single strand. This will be essentially the case in conductors that are in contact and are poorly insulated, except that at high frequencies the additional loss of energy due to heating of the imperfect contacts by the passage of the current from one strand to another may raise the resistance still higher.

Insulated Strands.—As the distance between the strands is increased, the resistance ratio falls, rapidly at first, and then more slowly toward the limit which holds for a single isolated strand. A very moderate thickness of insulation between the strands will quite materially reduce the resistance ratio, provided conduction in the dielectric is negligible.

Spiraling or twisting the strands has the effect of increasing the resistance ratio slightly, the distance between the strands being unchanged.

Transposition of the strands so that each takes up successively all possible positions in the cross section—as for example, by thorough braiding—reduces the resistance ratio but not as low as the value for a single strand.

Twisting together conductors, each of which is made up of a number of strands twisted together, the resulting composite conductor being twisted together with other similar composite conductors, etc., is a common method for transposing the strands in the cross section. Such conductors do not have a resistance ratio very much different from a simple bundle of well-insulated strands.

The most efficient method of transposition is to combine the strands in a hollow tube of basket weave. Such a conductor is naturally more costly than other forms of stranded conductor.

Effect of Number of Strands.—With respect to the choice of the number of strands, experiment shows that the absolute rise of the resistance in ohms depends on the diameter of a single strand, but is independent of the number of strands. Since, however, the direct-current resistance of the conductor is smaller the greater the number of the strands, the resistance ratio is greater the greater the number of strands. Reducing the diameter of the strands reduces the resistance ratio, the number of strands remaining unchanged, but to obtain a given current-carrying capacity, or a small enough total resistance, the total cross section must not be lowered below a certain limit, so that, in general, reducing the diameter of the strands means an increase in the number of strands.

With enameled strands of about 0.07 mm bare diameter twisted together to form a composite conductor the order of magnitude of the resistance ratio may be estimated by the following procedure. Calculate by Table 19, page 311, and Table 17, page 309, the resistance ratio for a single strand at the desired frequency (this value of R/R_0 will lie very close to unity), and carry out the same calculation for the equivalent solid wire, whose diameter will of course be $d\sqrt{n}$, where n = the number of strands and d = the diameter of a single strand. Then the resistance ratio for the stranded conductor will, for moderate frequencies, lie about one-quarter to one-third of the way between these two values, being closer to the lower limit. This holds for straight wires up to higher frequencies than for solenoids. (See critical frequency mentioned in second paragraph below.) Not all so-called litzendraht is as good as this by any means. For a woven tube the resistance ratio may be as low as one-tenth of the way from the lower to the upper limits mentioned.

Coils of Stranded Wire.—In the case of solenoids wound with stranded conductor, the resistance ratio is always larger than for the straight conductor, and at high frequencies may be two to three times as great. It is appreciably greater for a very short coil than for a long solenoid.

For moderate frequencies the resistance ratio is less than for a similar coil of solid wire of the same cross section as just stated, but for every stranded-conductor coil there is a critical frequency above which the stranded conductor has the larger resistance ratio. This critical frequency lies higher the finer the strands and the smaller their number. For 100 strands of say 0.07 mm diameter this limit lies above the more usual radio frequencies.

This supposes that losses in the dielectric are not important, which is the case for single-layer coils with strands well insulated. In multiple-layer coils of stranded wire, dielectric losses are not negligible at high frequencies.

77. TABLES FOR RESISTANCE CALCULATIONS

TABLE 17.—Ratio of High-Frequency Resistance to the Direct-Current Resistance

[See formulas (207), (208), and (209)]

x	$\frac{R}{R_0}$	Difference	x	$\frac{R}{R_0}$	Difference	x	$\frac{R}{R_0}$	Difference
0	1.0000	0.0003	5.2	2.114	0.070	14.0	5.209	0.177
0.5	1.0003	.0004	5.4	2.184	.070	14.5	5.386	.176
.6	1.0007	.0005	5.6	2.254	.070	15.0	5.562	.353
.7	1.0012	.0000	5.8	2.324	.070			
.8	1.0021	.0013	6.0	2.394	.069	16.0	5.915	0.353
.9	1.0034	.0018	6.2	2.463	.070	17.0	6.268	.353
						18.0	6.621	.353
1.0	1.005	0.003	6.4	2.533	0.070	19.0	6.974	.354
1.1	1.008	.003	6.6	2.603	.070	20.0	7.328	.353
1.2	1.011	.004	6.8	2.673	.070			
1.3	1.015	.005	7.0	2.743	.070	21.0	7.681	0.353
1.4	1.020	.006	7.2	2.813	.071	22.0	8.034	.353
1.5	1.026	.007	7.4	2.884	.070	23.0	8.387	.354
						24.0	8.741	.353
1.6	1.033	0.009	7.6	2.954	0.070	25.0	9.094	.353
1.7	1.042	.010	7.8	3.024	.070			
1.8	1.052	.012	8.0	3.094	.071	26.0	9.447	0.70
1.9	1.064	.014	8.2	3.165	.070	28.0	10.15	.71
2.0	1.078	.033	8.4	3.235	.071	30.0	10.86	.71
						32.0	11.57	.70
2.2	1.111	0.041	8.6	3.306	0.071	34.0	12.27	.71
2.4	1.152	.049	8.8	3.376	.070			
2.6	1.201	.056	9.0	3.446	.071	36.0	12.98	0.71
2.8	1.256	.062	9.2	3.517	.070	38.0	13.69	.71
3.0	1.318	.067	9.4	3.587	.071	40.0	14.40	.70
						42.0	15.10	.71
3.2	1.385	0.071	9.6	3.658	0.070	44.0	15.81	.71
3.4	1.456	.073	9.8	3.728	.071			
3.6	1.529	.074	10.0	3.799	.176	46.0	16.52	0.70
3.8	1.603	.075	10.5	3.975	.176	48.0	17.23	.71
4.0	1.678	.074	11.0	4.151	.176	50.0	17.93	3.54
						60.0	21.47	3.53
4.2	1.752	0.074	11.5	4.327	0.177	70.0	25.00	3.54
4.4	1.826	.073	12.0	4.504	.176			
4.6	1.899	.072	12.5	4.680	.176	80.0	28.54	3.53
4.8	1.971	.072	13.0	4.856	.177	90.0	32.07	3.54
5.0	2.043	.071	13.5	5.033	.176	100.0	35.61	....
						∞	∞	

TABLE 18.—Maximum Diameter of Wires for High-Frequency Resistance Ratio of 1.01

Frequency ÷ 10⁶	0.1	0.2	0.4	0.6	0.8	1.0	1.2	1.4	1.6	1.8	2.0	3.0
Wave length, meters	3000	1500	750	500	375	300	250	214.3	187.5	166.7	150	100
Material						Diameter in centimeters						
Copper	0.0956	0.0351	0.0177	0.0145	0.0125	0.0112	0.0102	0.0095	0.0089	0.0084	0.0079	0.0065
Silver	.0943	.0044	.0172	.0141	.0112	.0109	.0099	.0092	.0086	.0082	.0077	.0063
Gold	.0430	.0397	.0216	.0172	.0149	.8138	.0121	.0112	.0105	.0099	.0094	.0077
Platinum	.1120	.0793	.0560	.0457	.0396	.054	.0323	.0300	.0280	.0264	.0250	.0205
Mercury	.264	.187	.132	.1080	.0936	.0836	.0763	.0708	.0661	.0623	.0591	.0483
Manganin	.1794	.1361	.0962	.0729	.0631	.0564	.0515	.0477	.0446	.0420	.0399	.0325
Constantan	.1892	.1337	.0946	.0772	.0664	.0598	.0546	.0506	.0473	.0446	.0423	.0345
German silver	.1942	.1372	.0970	.0792	.0682	.0614	.0560	.0518	.0485	.0458	.0434	.0354
Graphite	.765	.541	.383	.312	.271	.243	.221	.204	.191	.180	.171	.140
Carbon	1.69	1.13	.801	.654	.566	.506	.462	.428	.400	.377	.358	.292
Iron μ=1000	0.00263	0.00186	0.00131	0.00108	0.00094	0.00083	0.00076	0.00070	0.00066	0.00062	0.00059	0.00048
μ=500	.00373	.00264	.00187	.00153	.00133	.00118	.00108	.00100	.00093	.00088	.00084	.00068
μ=100	.00838	.00590	.00418	.00340	.00295	.00264	.00241	.00223	.00209	.00197	.00186	.00152

TABLE 19.—Values of the Argument α_0 for Copper Wire 0.1 cm Diameter and Resistivity 1.724 Microhm-cms

f cycles per second	α_0	Difference	λ meters	f cycles per second	α_0	Difference	λ meters
100	0.1071	0.0443	....	50 000	2.395	0.229	6000
200	.1514	.0341	....	60 000	2.624	.210	5000
300	.1855	.0287	....	70 000	2.834	.195	4286
400	.2142	.0253	....	80 000	3.029	.184	3750
500	.2395	.0229	....	90 000	3.213	.174	3333
600	0.2624	0.0210	....	100 000	3.387	0.761	3000
700	.2834	.0195	....	150 000	4.148	.642	2000
800	.3029	.0184	....	200 000	4.790	.565	1500
900	.3213	.0174	....	250 000	5.355	.511	1200
1000	.3387	.1403	....	300 000	5.866	.818	1000
2000	0.4790	0.1076	....	333 333	6.184	0.380	900
3000	.5866	.0908	....	375 000	6.564	.452	800
4000	.6774	.0799	....	428 570	7.012	.561	700
5000	.7573	.0723	....	500 000	7.573	.723	600
6000	0.8296	0.0664	....	600 000	8.296	.664	500
7000	.8960	.0619	....	700 000	8.960	0.315	428
8000	.9579	.0581	....	750 000	9.275	.304	400
9000	1.0160	.055	....	800 000	9.579	.581	375
10 000	1.071	0.241	30 000	900 000	10.16	.55	333
15 000	1.312	.202	20 000	1 000 000	10.71	2.41	300
20 000	1.514	.341	15 000	1 500 000	13.12	5.43	200
30 000	1.855	.287	10 000	3 000 000	18.55	....	100
40 000	2.142	.253	7500				

TABLE 20.—Values of $\dfrac{R}{R_0}$ for Use with Formula (210)

β	$\dfrac{R}{R_0}$	Difference	β	$\dfrac{R}{R_0}$	Difference	β	$\dfrac{R}{R_0}$	Difference
0	1.000	....	1.0	1.086	0.037	2.5	2.477	0.111
0.1	1.000	....	1.1	1.123	.047	2.6	2.588	.109
.2	1.000	....	1.2	1.170	.059	2.7	2.697	.106
.3	1.001	....	1.3	1.229	.069	2.8	2.803	.104
.4	1.002	....	1.4	1.298	.080	2.9	2.907	.103
.5	1.006	0.002	1.5	1.378	.090	3.0	3.010	.101
0.55	1.008	.004	1.6	1.468	0.098	3.1	3.111	0.101
.60	1.012	.004	1.7	1.566	.106	3.2	3.212	.099
.65	1.016	.005	1.8	1.672	.111	3.3	3.311	.099
.70	1.021	.007	1.9	1.783	.115	3.4	3.410	.099
.75	1.028	.008	2.0	1.898	.117	3.5	3.509	.099
0.80	1.036	0.009	2.1	2.015	0.117	3.6	3.608	0.098
.85	1.045	.011	2.2	2.132	.117	3.7	3.706	.098
.90	1.057	.013	2.3	2.248	.115	3.8	3.804	.098
.95	1.070	.016	2.4	2.364	.113	3.9	3.902	.008
1.00	1.086	....	2.5	2.477	.111	4.0	4.600	....

MISCELLANEOUS FORMULAS AND DATA

78. WAVE LENGTH AND FREQUENCY OF RESONANCE

$$\lambda_{cm} = 1.8838 \times 10^{11} \sqrt{LC} \text{ (cgs electromagnetic units)} \tag{214}$$

$$= 6.283 \quad \sqrt{L \text{ cgs electromagnetic} \quad C \text{ cgs electrostatic}} \tag{215}$$

$$\lambda_m = 0.05957 \quad \sqrt{L \text{ cgs electromagnetic} \quad C \text{ micromicrofarad}} \tag{216}$$

$$= 1.884 \quad \sqrt{L \text{ microhenry} \quad C \text{ micromicrofarad}} \tag{217}$$

$$= 1884 \quad \sqrt{L \text{ microhenry} \quad C \text{ microfarad}} \tag{218}$$

$$= 59\,570 \quad \sqrt{L \text{ millihenry} \quad C \text{ microfarad}} \tag{219}$$

$$= 1\,884\,000 \sqrt{L \text{ henry} \quad C \text{ microfarad}} \tag{220}$$

$$f = \frac{159.2}{\sqrt{L \text{ henry} \quad C \text{ microfarad}}} \tag{221}$$

$$= \frac{5033}{\sqrt{L \text{ millihenry} \quad C \text{ microfarad}}} \tag{222}$$

$$= \frac{159\,200}{\sqrt{L \text{ microhenry} \quad C \text{ microfarad}}} \tag{223}$$

$$\omega = \frac{1000}{\sqrt{L \text{ henry} \quad C \text{ microfarad}}} \tag{224}$$

$$= \frac{31620}{\sqrt{L \text{ millihenry} \quad C \text{ microfarad}}} \tag{225}$$

$$= \frac{1\,000\,000}{\sqrt{L \text{ microhenry} \quad C \text{ microfarad}}} \tag{226}$$

$$T = \frac{1}{f} = \frac{2\pi}{\omega} \tag{227}$$

$$\lambda_m = \frac{2.998 \times 10^8}{f} \tag{228}$$

$$= \frac{1.884 \times 10^9}{\omega} \tag{229}$$

79. MISCELLANEOUS RADIO FORMULAS

When units are not specified, international electric units are to be understood. These are the ordinary units; based on the international ohm and ampere, the centimeter and the second. Full information is given on electric units in reference No. 152, Appendix 2.

Current in Simple Series Circuit.—

$$I = \frac{E}{\sqrt{R^2 + \left(\omega L - \frac{1}{\omega C}\right)^2}} \tag{230}$$

Phase Angle.—

$$\tan \theta = \frac{X}{R} = \frac{X_L - X_o}{R} \tag{231}$$

$$= \frac{\omega L - \frac{1}{\omega C}}{R} \text{ in simple series circuit.} \tag{232}$$

Sharpness of Resonance.—

$$\frac{\sqrt{\frac{I_r^2 - I_1^2}{I_1^2}}}{\frac{\pm (C_r - C)}{C}} = \frac{1}{R \omega C_r} = \frac{\omega L}{R} \tag{233}$$
(See p. 37.)

Current at Parallel Resonance.—

$$I = \frac{E R}{R^2 + \omega^2 L^2} \tag{234}$$
(See p. 39.)

Coefficient of Coupling.—

$$k = \frac{X_m}{\sqrt{X_1 X_2}} \tag{235}$$

$$= \frac{M}{\sqrt{L_1 L_2}} \text{ for direct and inductive coupling} \tag{236}$$

$$= \frac{\sqrt{C_1 C_2}}{C_m} \text{ for capacitive coupling.} \tag{237}$$
(See p. 49.)

Power Input in Condenser—

$$P = 0.5 \times 10^{-4} N C E_o^2 \text{ watts} \tag{238}$$

for C in microfarads, E_o in volts, and N = number of charges per second.

Power Loss in Condenser—

$$P = \omega C E^2 \sin \psi \tag{239}$$

Condenser Phase Difference—

$$\psi = r \omega C \tag{240}$$

for ψ in radians, r in ohms, C in farads.

$$\psi = 0.1079 \,\frac{rC}{\lambda} \text{ degrees} \tag{241}$$

for r in ohms, C in micromicrofarads, λ in meters.

$$\psi = 389 . \frac{rC}{\lambda} \text{ seconds} \tag{242}$$

for r in ohms, C in micromicrofarads, λ in meters.

$$r = \psi \times \frac{0.001}{C} \times \frac{\lambda}{1000} \times 0.154 \text{ ohms} \tag{243}$$

for ψ in minutes, C in microfarads, λ in meters.

Energy Associated with Inductance—

$$W = \frac{1}{2} L I^2 \tag{244}$$

Inductance of Coil Having Capacity:

$$L_a = \frac{L}{1 - \omega^2 CL} \tag{245}$$

for C in farads, L in the denominator in henries.

$$L_a = L \left(1 + 3.553 \,\frac{CL}{\lambda^2} \right) \text{ approximately} \tag{246}$$

for λ in meters, C in micromicrofarads, L in the parentheses in microhenries. This formula is accurate when the last term is small compared with unity.

Current Transformer—

$$\frac{I_1}{I_2} = \frac{n_2}{n_1} \left(1 + \frac{aR_2}{\omega L_2} \right) \tag{247}$$

(See p. 154.)

Audibility—

$$\frac{I}{I_t} = \frac{s + t}{s} \tag{248}$$

(See p. 166.)

Natural Oscillations of Horizontal Antenna.—

$$\lambda = \frac{1199}{m}\sqrt{C_0 L_0}, \quad m = 1, 3, 5, \ldots \quad (249)$$

for λ in meters, C_0 = capacity in microfarads for uniform voltage, L_0 = inductance in microhenries for uniform current.

Approximate Wave Length of Resonance for Loaded Antenna.—

$$\lambda = 1884\sqrt{C_0\left(L + \frac{L_0}{3}\right)} \quad (250)$$

where L = inductance of loading coil in microhenries and other quantities are as in preceding formula.

Radiation Resistance of an Antenna.—

$$R = 1580\left(\frac{h}{\lambda}\right)^2 \text{ ohms} \quad (251)$$

where h = height from ground to center of capacity, and h and λ are in the same units, and λ is considerably greater than the fundamental wave length.

Electron Flow From Hot Filament.—

$$I_s = AT^{\frac{3}{2}} e^{-\frac{b}{T}} \quad (252)$$

where I_s = electron current in milliamperes per centimeter^2 of filament surface, T = absolute temperature, and A and b depend on metal of filament; for tungsten $A = 2.5 \times 10^{10}$, $b = 52500$.

Electron Current in 3-Electrode Tube.—

$$I_s = k \ (E_s + k_1 v_1)^2 \quad (253)$$

where E_B = plate voltage, v_1 = grid voltage, k_1 = amplification constant.

Resistance Measurement by Resistance—Variation Method Using Undamped Emf.—

$$R = R_1 \frac{I_1}{I - I_1} \quad (254)$$

Resistance Measurement by Resistance—Variation Method Using Impulse Excitation.—

$$R = R_1 \frac{I_1^2}{I - I_1^2} \quad (255)$$

Resistance Measurement by Reactance-Variation Method Using Undamped emf.—

$$R = X_1 \sqrt{\frac{I_1^2}{I_2^2 - I_1^2}} \quad (256)$$

where X_1 = change of reactance between the two observations of current. Various particular cases of this formula are given in section 50.

Natural Frequency of Simple Series Circuit.—

$$f = \frac{1}{2\pi}\sqrt{\frac{1}{CL} - \frac{R^2}{4L^2}} \tag{257}$$

$$\omega = \frac{1}{\sqrt{CL}\sqrt{1 + \left(\frac{\delta}{2\pi}\right)^2}} \tag{258}$$

Number of Oscillations to Reduce Current to 1 Per Cent of Initial Value in Wave Train.—

$$n = \frac{4.6}{\delta} \tag{259}$$

Logarithmic Decrement.—

$$\delta = \log_{\epsilon}\frac{I_1}{I_2} = \frac{a}{f} \tag{260}$$

$$= \pi\frac{R}{\omega L} = \pi R\omega C = \pi R\sqrt{\frac{C}{L}}$$

$$= \frac{\pi}{\text{sharpness of resonance}}$$

$= \pi \times$ phase difference of condenser or coil, the resistance being in one or the other

$$= \frac{\text{average energy dissipated per cycle}}{2 \times \text{average magnetic energy at the current maxima}}$$

$$\delta = 0.00167\,\frac{R\lambda}{L} \tag{261}$$

for R in ohms, λ in meters, L in microhenries.

$$\delta = 5918\,\frac{RC}{\lambda} \tag{262}$$

for R in ohms, λ in meters, C in microfarads.

$$\delta = 3.1416\,R\sqrt{\frac{C}{L}} \tag{263}$$

for R in ohms, C in microfarads, L in microhenries.

Current at resonance Produced by Slightly Damped emf Induced in a Circuit.—

$$I^2 = \frac{N\,E_o^2}{16f^2L^2\delta'\delta\,(\delta' + \delta)} \tag{264}$$

Decrement Measurement by Reactance—Variation Method.—

$$\delta' + \delta = \pi\frac{C_2 - C_1}{C_2 + C_1}\sqrt{\frac{I_1^2}{I_r^2 - I_1^2}} \tag{265}$$

(See p. 187 for variations of this formula.)

80. PROPERTIES OF METALS

TABLE 21

Metal	Microhm-centimeters at 20° C	Temperature coefficient at 20° C	Specific gravity	Tensile strength, lbs./in.²	Melting point, °C
Advance. See Constantan.					
Aluminum	2.828	0.0039	2.70	30 000	659
Antimony	41.7	.0036	6.6	....	630
Bismuth	120	.004	9.8	....	271
Brass	7	.002	8.6	70 000	900
Cadmium	7.6	.0038	8.6	....	321
Calido. See Nichrome.					
Climax	87	.0007	8.1	150 000	1250
Constantan	49	.00001	8.9	120 000	1190
Copper, annealed	1.7241	.00393	8.89	30 000	1083
Copper, hard-drawn	1.771	.00382	8.89	60 000	....
Eureka. See Constantan.					
Excello	92	.00016	8.9	95 000	1500
German silver, 18 per cent	33	.0004	8.4	150 000	1100
German silver, 30 per cent. See Constantan.					
Gold	2.44	.00342	19.3	20 000	1063
Ia Ia. See Constantan.					
Ideal. See Constantan.					
Iron, 99.98 per cent pure	10	.0050	7.8	....	1530
Iron. See Steel.					
Lead	22	.0039	11.4	3 000	327
Magnesium	4.6	.004	1.74	33 000	651
Manganin	44	.00001	8.4	150 000	910
Mercury	95.783	.00089	13.546	0	−38.9
Molybdenum, drawn	5.7	.004	9.0	....	2500
Monel metal	42	.0020	8.9	160 000	1300
Nichrome	100	.0004	8.2	150 000	1500
Nickel	7.8	.006	8.9	120 000	1452
Palladium	11	.0033	12.2	39 000	1550
Phosphor bronze	7.8	.0018	8.9	25 000	750
Platinum	10	.003	21.4	50 000	1755
Silver	1.59	.0038	10.5	42 000	960
Steel, E. B. B.	10.4	.005	7.7	53 000	1510
Steel, B. B.	11.9	.004	7.7	58 000	1510
Steel, Siemens-Martin	18	.008	7.7	100 000	1510
Steel, manganese	70	.001	7.5	230 000	1260
Superior. See Climax.					
Tantalum	15.5	.0031	16.6	....	2850
Therlo	47	.00001	8.2	....	....
Tin	11.5	.0042	7.3	4000	232
Tungsten, drawn	5.6	.0045	19	500 000	3000
Zinc	5.8	.0037	7.1	10 000	419

The resistivities given in Table 21 are values of ρ in the equation $R_0 = \rho \frac{l}{s}$, where l = length in centimeters and s = cross section in square centimeters. This formula gives the low-frequency or direct-current resistance of a conductor. For the calculation of resistances at high frequencies, see Tables 17 to 20, pages 309–311.

The values given for resistivity and temperature coefficient of copper are the international standard values for commercial copper. Any departure from this resistivity is accompanied by an inverse variation in the temperature coefficient. This is true in a general way for other metal elements. In the case of copper the resistivity and temperature coefficient are inversely proportional, to a high degree of accuracy.

The "temperature coefficient at 20°C" is a_{20} in the equation $R_t = R_{20} (1 + a_{20}[t - 20])$. In some cases the temperature variation does not follow a straight-line law; in such cases a_{20} applies only to a small range of temperature close to 20°. Steel is an example, the resistance rise at high temperatures being faster than proportional to temperature.

Constantan and the other wires (Advance, etc.) having substantially the same properties, are alloys of approximately 60 per cent copper and 40 per cent nickel. They are used in rheostats and measuring instruments.

German silver is an alloy of copper, nickel, and zinc. The per cent stated indicates the percentage of nickel.

Manganin contains about 84 per cent copper, 12 per cent manganese, and 4 per cent nickel. It is the usual material in resistance coils. Its very small thermal electromotive force against copper is one of its main advantages. The similar alloy, therlo, is used for the same purposes.

Monel metal is an alloy containing approximately 71 per cent nickel, 27 per cent copper, and 2 per cent iron.

FIG. 214.—*Variable condensers used as standards of capacity*

FIG. 215.—*Single-layer coils used as standards of inductance*

FIG. 216.—*Multiple-layer standard coil*

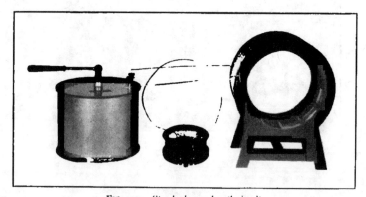

FIG. 217.—*Standard wave length circuit*

APPENDIXES

❧

APPENDIX 1.—RADIO WORK OF THE BUREAU OF STANDARDS

The functions of the radio laboratory of this Bureau include the maintenance of standards for radio measurements, the testing of instruments and apparatus, technical assistance in radio matters to various branches of the Government, and researches in the theory and practice of radio communication. The activities of the Bureau in some of these lines have been to a considerable extent covered in the foregoing sections. A more comprehensive account is given here of the facilities, accomplishments, and aims of this laboratory.

This account does not include a description of the work of the United States naval radiotelegraphic laboratory or of the Signal Corps laboratory, both of which are located at the Bureau of Standards. They were in existence before the Bureau's own radio laboratory was established, and the publications of the Naval Laboratory are printed in the Bulletin of the Bureau of Standards. A list of these publications is given in the "Bibliography," page 329.

1. DEVELOPMENT AND MAINTENANCE OF STANDARDS

Capacity.—The qualities desirable in a condenser to be used as a standard at radio frequencies are: Constancy of capacity with varying frequency and temperature and other conditions, small resistance or phase difference, careful shielding, and convenience of design. The quartz-pillar air condensers described above (p. 120) have these qualities and are satisfactory fundamental standards for radio measurements. They are the result of many years' experience at this Bureau in the measurement and design of condensers. The laboratory has a set of variable condensers of this type, with maximum capacities ranging from 0.0001 to 0.0075 microfarad. Having continuously variable capacity, they are very convenient to use in a standard circuit. Fixed-value condensers of the same general type with greater capacity are also used. Good mica condensers, well made and properly shielded, may also be used as standards at radio frequencies. The best mica condensers have lower phase differences than many air condensers of ordinary design, because of the solid dielectric used to insulate the plates in the air condensers. Fixed-value condensers for radio use should preferably be independent and not parts of a permanently connected set of condensers, on account of mutual capacities between the parts of such a set.

The capacities of air condensers used as radio standards are determined by low-frequency measurements, either by the absolute Maxwell bridge method [36] at a frequency of 100 per second, or by alternating-current comparison [37] with standard condensers at frequencies from 100 to 3000. The plate and lead resistances and inductances of these condensers are negligibly small, the insulation resistance is extremely high, and the phase difference due to absorption is very small. A few of the condensers have a phase difference which can just be detected at low settings. It is so small as to cause no change of capacity, as shown by agreement of the capacities at low and high settings at different frequencies. These condensers have practically zero temperature coefficient and have remained constant in capacity.

---

[36] See reference No. 174, Appendix 2.   [37] See reference No. 176, Appendix 2.

Inductance.—The problem of developing standards of inductance for use at radio frequencies is mainly that of minimizing resistance and distributed capacity. These requirements are both met by the single-layer coil for inductances of moderate value. The shape of coil having the minimum length of wire (and hence minimum resistance if the cross section of the wire is specified) for a required inductance is given on pages 286 to 292. The capacity of a coil is roughly proportional to the radius of the coil and independent of the number of turns and length. For a single-layer coil having a close winding the value in micromicrofarads is approximately equal to the numerical value of the diameter in centimeters. For a single-layer coil with spaced winding on an open form, like those described below, the capacity in micromicrofarads is approximately equal to the numerical value of the radius in centimeters. Hence a coil should be made longer and of smaller diameter than the theoretical shape indicated for minimum resistance in order to reduce the capacity.

The single-layer coils used as standards in this laboratory conform to these principles. The set shown in Fig. 215, facing page 318, ranges in diameter from 13 to 38 cm and in inductance from 60 to 5350 microhenries. The capacities of the coils with their leads range from 9 micromicrofarads for the smallest to 16 for the largest. The capacities are kept small by eliminating as much dielectric as possible from the neighborhood of the wire. The coils are wound with silk-covered "litzendraht", with the turns spaced. The open form gives the coils the shape of a 12-sided polygon instead of a circle.

Multiple-layer coils are used as standards for inductances larger than any of these single-layer standards. These are satisfactory if the wires are spaced well apart and the amount of dielectric between the turns and layers is kept small. On this account it is not desirable to impregnate such coils with insulating compound. Such a coil is shown in Fig. 216, facing page 319.

The inductances of the standard coils are determined by intercomparison in circuits using the standard air condensers referred to above. The basis of these intercomparisons is a rectangular inductance consisting of a single turn of copper tubing. The inductance of this, obtained by calculation, is 9 microhenries. For discussion of these determinations see page 247.

Wave Length.—The wave-length standards consist essentially of standard circuits made up of the standard condensers and inductance coils just described. As shown in Fig. 217, facing page 313, the circuit includes a pair of leads; the inductance coil is considered to include these when its value is determined. A wire leading to ground is connected to the shielded side of the condenser. A current indication is obtained in a thermal ammeter in a separate circuit near the standard circuit. This separate circuit is placed by trial at such a distance that it does not affect appreciably the capacity or inductance of the standard circuit. An alternative method of observing the current in the standard circuit is the use of a thermoelement in series, the circuit being standardized with the thermoelement in.

The range of wave lengths obtained with the coils and condensers described above is from 100 to 13 000 meters.

The work of the Bureau in connection with wave-length standardization includes also the development of the decremeter described on pages 196–199. This instrument is a wave meter which has a more nearly uniform scale of wave lengths than wave meters employing the ordinary condenser with semicircular plates. It is built for wave lengths from 75 meters up.

Current.—High-frequency current standardization is at present based upon thermal ammeters. For small currents the standard instruments are thermoelements. These are made of such fine wire that the resistance does not change with frequency. When made with a resistance of 1 ohm or less, a thermoelement may be inserted directly in a radio circuit without reducing the current materially.

FIG. 218.—*Interior view of decremeter*

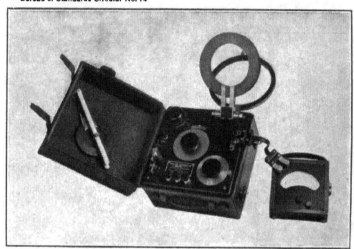

FIG. 219.—*Small-type decremeter*

FIG. 220.—*Navy-type decremeter*

For currents of intermediate value, a hot-wire ammeter with a single wire is taken as a standard. The principal precaution necessary is that the heated wire be fine enough to remain constant in resistance at all frequencies used.

For measurements of large currents, instruments with multiple wires or strips are taken as standards, in which careful investigation has shown that the current distribution among the wires or strips does not change with frequency. Ammeters of the cylindrical type with a thermocouple on each wire or strip have been developed for this purpose. The questions of errors and design are treated above, section 41. The standard instruments thus far developed are suitable for measuring currents up to 50 amperes at frequencies up to 1 000 000.

Resistance and decrement.—The standards used in the determination of resistance and decrement are of two classes. In the first class, in which resistance is measured by the substitution or the deflection method, resistance standards of manganin wire are used. These are short, straight lengths of fine wire, the substitution of which in a circuit does not appreciably change its inductance. (See p. 178–180 for further details.)

In the second class of measurement, the determination of resistance depends ultimately on the reactance of a circuit and the deflections of an ammeter. From the variation of reactance required to produce a certain change of current in the circuit the resistance is obtained. In the decremeter described on page 196, which was developed at this Bureau, the decrement is obtained directly by manipulation of the instrument without the necessity of any calculation. A dial is graduated in terms of decrements from 0 to 0.3 readable to 0.001. Photographs of several types of the instrument are given in Figs. 218 to 220.

2. TESTING OF INSTRUMENTS AND MATERIALS

Most of the radio apparatus which the Bureau is called on to test is standardized by direct comparison with the standards described in the preceding section. The fees which have been established for testing radio and other electrical apparatus, and instructions to applicants for tests, are given in this Bureau's Circular No. 6, Fees for Electric, Magnetic, and Photometric Testing. For tests not listed there, a special fee of nominal amount is charged. Unless otherwise specified, apparatus is tested with undamped current using a pliotron as the source. Such current is very steady and gives the maximum accuracy of measurement. In general, the Bureau does not certify an accuracy better than 1 per cent on any radio apparatus.

Wave meters.—A wave meter is tested by direct comparison with a standard circuit, both being coupled to the same source of high-frequency current. For the procedure see section 30. If a ground connection is to be used on the wave meter, it is tested with a ground on. It is usually most convenient to use an ordinary commercial wave meter without a ground connection. Grounding makes very little difference in the indications of the instrument, except in wave meters where the condenser has unusually small capacity at the low settings. Wave meters which are to be used as instruments of precision are tested at the points specified by the applicant for test.

Coils.—An inductance coil is tested by a substitution method, other coils of nearly the same value being substituted in a circuit which is tuned to resonance by a variable condenser. By varying the capacity of the condenser, or if necessary inserting additional inductance, the wave length is varied, and a curve of inductance against wave length may be plotted. Such a curve is shown on page 64.

Condensers.—Condensers are also tested by substitution, either by placing the test condenser and the standard successively in the same position or by using a double-throw switch. The use of variable standard condensers makes this measurement very simple.

Ammeters.—High-frequency ammeters are standardized by comparison with a standard ammeter in series in the same circuit. Test is usually made at more than one wave length.   No regular fees have been established for this as yet, each test being subject to a special fee depending on the time consumed in the measurement.

Resistance Measurements.—The resistances of high-frequency resistance standards, of wires or other conductors, and of coils, condensers, or circuits, are measured in any of the ways mentioned in sections 47 to 50 above.   These include substitution or deflection methods in terms of standards, and the reactance-variation methods.   In the case of stranded wire submitted for high-frequency resistance measurement, it is desirable to measure also the low-frequency or direct-current resistance of each separate strand and the insulation resistance between strands.

Insulating Materials.—Tests are made of the dielectric loss or phase difference of insulating materials if submitted in large thin sheets.   Measurements of dielectric strength are made with low-frequency voltage up to 100 000 volts.   An equipment has also been developed for voltage tests at radio frequencies up to 20 000 volts.

Operating Apparatus.—Complete transmitting and receiving sets, accessories, and parts of sets are tested when the circumstances render the test of such importance as to justify the Bureau in undertaking the work.   Tests of the performance of complete sets have not yet been standardized, as each set submitted presents a distinct problem.   Such a test may include: Output of transmitter, wave forms of current and voltage of power supply circuits, purity and decrement of generated wave, wave lengths of transmitter and receiver, selectivity and sensitivity of receiver.

3. RADIO ENGINEERING FOR THE GOVERNMENT

The testing and research work is of direct value to the Army, Navy, and various other branches of the Government, but in addition to this the laboratory performs special services for Government Bureaus, in particular those of the Department of Commerce.   Some of the special lines of work thus pursued are described below. Technical information is also furnished upon request.   The subjects upon which information has thus been furnished include: The installing of transmitting and receiving equipment, the efficiency of radio apparatus, the adjustment of equipment to comply with the law, the design of measuring instruments, formulas, and data. Assistance is rendered the Government in the preparation of legislation on radio matters.

Design of Instruments.—Portable testing equipments have been developed for the radio inspectors of the Bureau of Navigation of the Department of Commerce.   The decremeter and the voltammeter for this purpose (described in preceding sections) were designed, construction supervised, and calibrated here.   Technical problems in connection with instruments, which have arisen in the radio inspections, have been referred to this laboratory for solution.

Design of Radio Sets.—Complete radio transmitting and receiving sets have been designed and furnished to three of the Bureaus of the Department of Commerce. These are in use on the ships of the Lighthouse Service, Bureau of Navigation, and Coast and Geodetic Survey.   The transmitters are built in compact panel form and are supplied with 1 kw of power in a motor generator delivering 500-cycle current. This current flows in a closed-core transformer, adjusted for maximum efficiency, to the secondary of which are connected a quenched gap of special design, mica condensers, and a flat spiral coil.   A simple switch sets the wave length on 600, 750, and 1000 meters.   The sets handle relatively little traffic, and have a range of about 200 km.   Two views of the transmitter are shown in Figs. 221 and 222, facing page 322. The transformer and the inductance spirals, which were given special attention in the development work, are shown in Fig. 223, facing page 323.

The receiver designed for these sets consists of two circuits, the antenna circuit and a closed detecting and measuring circuit inductively coupled to it.   The closed circuit and the antenna loading coils and variable condenser are all contained in a

FIG. 221.—*Transmitting set designed by Bureau of Standards (front view)*

FIG. 222.—*Transmitting set designed by Bureau of Standards (rear view)*

FIG. 175.—*Quenched gap plate showing the circular silver sparking surface*

FIG. 223.—*Inductance spirals and transformer used in the transmitting set shown in Figs. 221 and 222*

FIG. 224.—*Receiving set designed by Bureau of Standards*

compact cabinet. The closed circuit includes a variable condenser of the decremeter type, and serves as a wave meter and decremeter as well as acting as a receiver by virtue of the crystal detector connected across the condenser. The receiver may be tuned to wave lengths from about 500 to 2500 meters. Two views are shown in Fig. 224, facing page 323.

Fog Signaling Apparatus.—The Bureau of Standards has been active in its efforts to promote safety at sea by means of radio apparatus. An equipment was designed and constructed for use at a lighthouse, which should efficiently supplement the light of a lighthouse during fog and prove of great assistance to navigation. An automatic transmitting device is arranged to send out a characteristic signal once every minute on a short wave length, so that it will be readily received by all ships within a few miles of the lighthouse. A direction finder was developed for use on ships receiving the signal, so that they can get their bearings by radio.

Field Work.—Inspection and other trips are made at the request of other Government bureaus. Assistance has thus been rendered to the Bureau of Navigation of the Department of Commerce in order to solve technical problems that have arisen in the radio inspections. Such problems have included the equipment of emergency radio sets on shipboard, cases of interference, use of instruments and testing equipment, etc.

4. RESEARCH WORK.

Military Researches.—The testing of instruments and materials is of direct or indirect benefit to the military departments. Additional service of military value is being rendered by the laboratory through the results of most of the investigations which are in progress. These investigations are of both a scientific and engineering nature. It is obviously impossible to publish any description of this work.

Radio Instruments and Methods of Measurement.—A number of problems in radio measurements are being studied in the laboratories of the Bureau of Standards. Some of these have been brought to the point where a publication has been issued or a testing routine established, but all of them remain fruitful fields for investigation. Among the more important problems is that of establishing wave-length standards. The standard circuits which have been developed are described above. The production and measurement of large currents and high voltages is another branch of the work. In this connection one publication has been issued, Scientific Paper No. 206, "High-Frequency Ammeters," and a special type of volt-ammeter has been designed. These investigations have shown that simplicity of circuit is a great desideratum for many radio measurements.

The measurement of resistance and decrement has received considerable attention. A number of methods have been used, and their limitations studied. An apparatus for quick measurements has been developed; it is described in Scientific Paper No. 235, "A Direct-Reading Decremeter for Measuring the Logarithmic Decrement and Wave Length of Electromagnetic Waves."

Properties of Conductors and Insulators.—Data are obtained on the ratio of high-frequency to low-frequency resistance of stranded wire of various kinds. This work may be extended to strips, tubes, and other special forms of conductors. Insulating materials are studied for dielectric loss, dielectric constant and its variation with frequency, surface flashover voltage, etc. There is great need for systematic study both of the methods of measurement and of the properties of these materials.

Inductance Coils.—The capacity and the resistance of radio coils and their effect upon the inductance furnish an interesting problem. The effects of varying shape, size, pitch and size and kind of conductor, insulation of conductor, and material and kind of mounting, all require investigation, as well as the modes of connection to radio coils and the effects produced by combinations of coils.

Electron Tubes.—The characteristics and applications of three-electrode thermionic tubes constitute a most important field of investigation. These tubes have been found to be excellent sources of current for laboratory measurements. A number of applications to military uses are under development. The characteristic curves of tubes are studied, and different types of tubes compared as amplifiers, generators, and detectors. Special attention is given to the production of maximum current in generating circuits for particular purposes, and to the modulation of the radio-frequency current.

Antennas.—Some of the great variety of problems presented by the antenna are under study. The properties, functioning, and merits of antennas of various forms for particular purposes are investigated. The means of supplying current to the antenna are studied. The investigation includes the consideration of the behavior and transmission of the electromagnetic waves emitted from an antenna. Measurements of antenna resistance, inductance, and capacity are made. One publication has been issued, Scientific Paper No. 269, "Effect of Imperfect Dielectrics in the Field of a Radiotelegraphic Antenna."

APPENDIX 2.—BIBLIOGRAPHY

This bibliography is by no means comprehensive. A few of the more important references are given for each of the subjects treated in the text. In many of the publications listed here references are given to previous publications. Bibliographies of the current literature have been given bimonthly in the "Jahrbuch der drahtlosen Telegraphie" since 1907. Articles on radio measurements as well as other phases of radio communication appear in the bimonthly "Proceedings of the Institute of Radio Engineers."

ELEMENTARY ELECTRICITY.

1. Elements of Electricity and Magnetism, J. J. Thomson; 4th ed., 1909 (Cambridge).
2. Modern Views of Electricity, O. J. Lodge; 1889 (MacMillan).
3. The Elements of Physics, Vol. II, Electricity and Magnetism, Nichols and Franklin; 1905 (MacMillan).
4. Electricity and Magnetism, R. T. Glazebrook; 1910 (Cambridge).
5. Elements of Electricity for Technical Students, W. H. Timbie; 1911 (John Wiley & Sons).
6. Magnetism and Electricity for Students, H. E. Hadley; 1910 (MacMillan).
7. The Elements of Electricity and Magnetism, Franklin and MacNut; 1914 (Mac-Millan).
8. Elementary Lessons in Electricity and Magnetism, S. P. Thompson; 7th ed., 1915 (MacMillan).
9. A Treatise on Electricity, F. B. Pidduck; 1916 (Cambridge).
9a. Electricity and Magnetism, S. G. Starling; 1912 (Longmans, Green & Co.).

ATLERNATING CURRENTS.

11. Alternating Currents, Bedell and Crehore; 4th ed., 1901 (McGraw-Hill).
12. Alternating Currents and Alternating Current Machinery, D. C. and J. P. Jackson; 1896 (MacMillan).
13. The Theory of Alternating Currents (2 vols.), A. Russell; 2d ed., 1914 (Cambridge).
14. Kapazität und Induktivität, E. Orlich; 1909.
15. Calculation of Alternating Current Problems, L. Cohen; 1913 (McGraw-Hill).
16. The Foundations of Alternating Current Theory, C. V. Drysdale; 1910 (E. Arnold).
17. Transient Electric Phenomena and Oscillations, C. P. Steinmetz; 1909 (McGraw-Hill).

COUPLED CIRCUITS.

21. Currents in Coupled Circuits; A. Oberbeck; Annalen der Physik, **291**, p. 623; 1895.
22. Use of Coupled Circuits; F. Braun; Physikalische Zs., **8**, p. 148; 1901.
23. Coupling phenomena; M. Wien; Annalen der Physik, **61**, p. 151, 1897; **25**, p. 1, 1908.
24. Maximum Current in the Secondary of a Transformer; J. S. Stone; Physical Review, **32**, p. 399; 1911.
25. Cisoidal Oscillations; G. A. Campbell; Trans. A. I. E. E., **30**, p. 873; 1911.
26. The Impedances, Angular Velocities, and Frequencies of Oscillating-Current Circuits; A. E. Kennelly; Proc. I. R. E. **4**, p. 47; 1916.
27. Alternating and Transient Currents in Coupled Electrical Circuits; F. E. Pernot; University of California, publications in Engineering, **1**, p. 161; 1916.
28. Oscillograph Demonstrations of Coupled Circuits; G. W. O. Howe; Proc. Physical Society London, **23**, p. 237; 1911. J. A. Fleming; Proc. Physical Society London, **25**, p. 217; 1913.
29. Mechanical Models; T. R. Lyle; Phil. Mag., **25**, p. 567; 1913. W. Deutsch; Physikalische Zs., **16**, p. 138; 1915.

ANTENNA CALCULATIONS.

31. Theory of Horizontal Antennas; J. S. Stone; Trans. Int. Elec. Congress, St. Louis, **3**, p. 555; 1904.
32. Theory of Loaded Antenna; A. Guyau; La Lumière Electrique, **15**, p. 13; 1911.
33. Capacity of Radiotelegraphic Antennas; G. W. O. Howe; Electrician, **73**, pp. 829, 859, 906, 1914; **75**, p. 870; 1915.
34. The Electrical Constants of Antennas; L. Cohen; Elec. World, **65**, p. 286; 1915.

DAMPING.

41. Theory of Free Oscillations; Alternating Current Phenomena, C. P. Steinmetz; Appendix II, p. 709; 4th ed., 1908.
42. Decrements in Coupled Circuits; V. Bjerknes; Annalen der Physik, **44**, pp. 74, 92, 1891; **291**, p. 121, 1895. M. Wien, Annalen der Physik, **25**, p. 625, 1908; **29**, p. 679, 1909.
43. Linear Decrement: J. S. Stone; Electrician, **73**, p. 926; 1914. Proc. I. R. E., **2**, p. 307, 1914; **4**, p. 463, 1916.

ELECTROMAGNETIC WAVES.

51. A Treatise on Electricity and Magnetism; J. C. Maxwell; 1873.
52. Recent Researches in Electricity and Magnetism; J. J. Thomson; 1893.
53. Electromagnetic Theory (3 vols.); O. Heaviside; 1893.
54. Signaling Through Space Without Wires; O. J. Lodge; 1894.
55. Derivation of Equations of a Plane Electromagnetic Wave; E. B. Rosa; Phys. Rev., **8**, p. 282; 1899.
56. Electric Waves; H. Hertz (translated into English by D. E. Jones); 1900.
57. Maxwell's Theory and Wireless Telegraphy; H. Poincaré (translated into English by F. K. Vreeland); 1904.
58. Researches in Radiotelegraphy; J. A. Fleming; Smithsonian Report for 1909, p. 157.

RADIO MEASUREMENTS AND MISCELLANEOUS.

61. The Principles of Electric Wave Telegraphy and Telephony; J. A. Fleming; 3d ed., 1916.
62. Les Oscillations Electriques; C. Tissot; 1910.
63. Radiotelegraphisches Praktikum; H. Rein; 1912.

64. Wireless Telegraphy; J. Zenneck (translated into English by A. E. Seelig); 1915.
65. Wireless Telegraphy and Telephony, A Handbook; W. H. Eccles; 1916.
66. Radio Communication; J. Mills; 1917.
67. Standardization Rules, Institute of Radio Engineers; 1915.

WAVE LENGTH.

71. Die Frequenzmesser und Dämpfungsmesser der drahtlosen Telegraphie; E. Nesper; 1907.
72. Standard Wave Length Circuits; A. Campbell; Phil. Mag., **18**, p. 794; 1909. Electrician, **64**, p. 612; 1910.
73. Calibration of Wavemeters; G. W. O. Howe; Electrician, **69**, p. 490; 1912.
74. Wavemeter Standardization; Diesselhorst; Elektrotechnische Zs., **29**, p. 703; 1908.
75. Pointer-Type Wavemeter; Ferrié and Carpentier; Jahrb. d. drahtl. Tel., **5**, p. 106; 1911.
76. Practical Uses of the Wavemeter in Wireless Telegraphy; J. O. Mauborgne; 1914.
77. Oval Diagram for Wave Length Calculations; W. H. Eccles; Electrician, **76**, p. 388; 1915.

CAPACITY.

81. Square-Plate Condenser for Uniform Scale of Wave Lengths; C. Tissot; Journal de Physique, **2**, p. 719; 1912.
82. Rotary Condenser for Uniform Scale of Wave Length; W. Duddell; Jour. I. E. E., **52**, p. 275; 1914.
83. A.-c. Resistance of Condensers; Fleming and Dyke; Electrician, **68**, pp. 1017, 1060, 1912; **69**, p. 10, 1912. G. E. Bairsto; Electrician, **76**, p. 53, 1915.
84. Calculation of Capacity Using Method of Images; "Alternating Currents"; A. Russell; Vol. I, chaps. 5 and 6; 1914.

INDUCTANCE.

91. The Effects of Distributed Capacity of Coils Used in Radiotelegraphic Circuits; F. A. Kolster; Proc. I. R. E., **1**, p. 19; 1913.
92. Distributed Capacity of Single-Layer Solenoids; J. C. Hubbard; Phys. Review, **9**, p. 529; 1917.
93. Development of Inductance Formulas; "Alternating Currents"; A. Russell; Vol. I, chaps. 2 and 3; 1914. "Absolute Measurements in Electricity and Magnetism"; A. Gray; Vol. II, part 1, chap. 6.

CURRENT MEASUREMENT.

101. Thermoelements for High-Frequency Measurements; Dowse; Electrician, **65**, p. 765; 1910.
102. Hot-Strip Ammeters for Large High-Frequency Currents; R. Hartmann-Kempf; Elektrotechnische Zs., **32**, p. 1134; 1911. G. Eichhorn; Jahrbuch d. drahtl. Tel., **5**, p. 517; 1912.
103. High-Frequency Current Transformer; Campbell and Dye; Proc. Royal Soc., **90**, p. 621; 1914.
104. Use of Iron in High-Frequency Current Transformer; McLachlan; Electrician, **78**, p. 382; 1916.
105. Use of Galvanometer in Audion Plate Circuit; L. E. Whittemore; Phys. Review, **9**, p. 434; 1917.
106. Measurement of Signal Intensity with Crystal Detector; J. L. Hogan; (Marconi) Year-Book of Wireless Telegraphy, p. 662; 1916.

107. Measurements With Crystal and Telephone; J. Zenneck; Proc. I. R. E., **4**, p. 363; 1916.
108. Current Measurement With the Audion; L. W. Austin; Jour. Wash. Acad. Sciences, **6**, p. 81; 1916. Proc. I. R. E., **4**, p. 251; 1916. Electrician, **78**, p. 465; 1917. Proc. I. R. E., **5**, p. 239; 1917.

HIGH-FREQUENCY RESISTANCE.

111. Skin Effect in Round Wires; Lord Rayleigh; Phil. Mag., pp. 382, 469; 1886; Sci. Papers, Vol. II, pp. 486, 495. Skin Effect in Round Wires; Lord Kelvin; Math. and Phys. Papers, Vol. III, p. 491; 1889.
112. Skin Effect in Stranded Conductors to Oscillatory Currents; F. Dolezalek; Ann. der Phys., (4), **12**, p. 1142; 1903.
113. Passage of High-Frequency Current Through Coils; M. Wien; Ann. der Phys., (4), **14**, p. 1; 1904.
114. Long Solenoids at High Frequencies, Mathematical Theory; A. Sommerfeld; Ann. der Phys., (4), **15**, p. 673, 1904; (4), **24**, p. 609, 1907.
115. Calorimetric Measurements of High-Frequency Resistance of Solenoids; T. Black; Ann. der Phys., **19**, p. 157; 1906.
116. Measurements on Stranded Conductors; R. Lindemann; Verh. deutsch. Phys. Gesel., **11**, p. 682; 1909.
117. Theory for Stranded-Conductor Solenoids; Möller; Ann. der Phys., **36**, p. 738, 1911; and Jahr. draht. Tel., **9**, p. 32, 1914.
118. Measurements on Single and Multiple Layer Coils; Esau; Ann. der Phys., **34**, p. 57; 1911.
119. Skin Effect in Flat Coils and Short Cylindrical Coils; Lindemann and Hüter; Verh. deutsch. Phys. Ges., **15**, p. 219; 1913.
120. The Alternating-Current Resistance of Long Coils of Stranded Wire, Theory; Rogowski; Arch. f. Elect., **8**, p. 264; 1915.
121. Bibliography, and Measurements on Wires and Strips; Kennelly, Laws, and Pierce; Proc. A. I. E. E., **34**, p. 1749; 1915.
122. Bibliography, and Measurements on Solid and Stranded Conductors; Kennelly and Affel; Proc. I. R. E., **4**, p. 523; 1916.
123. High-Frequency Resistance of Multiply-Stranded Insulated Wire; G. W. O. Howe; Proc. Royal Society London, **93**, p. 468; 1917.
124. The Accuracy of High-Frequency Resistance Measurements; S. Loewe, Jahrbuch d. Drahtlosen Telegraphie, **7**, p. 365; 1913.

ELECTRON TUBES.

131. Theory of Thermionic Emission; O. W. Richardson; Phil. Trans., **202**, p. 516; 1903.
132. Audion Detector and Amplifier; L. De Forest; Electrician, **78**, p. 842; 1914. Elec. World, **65**, p. 465; 1914.
133. Theory of Electron Tubes; I. Langmuir; Phys. Review, **2**, p. 450; 1913. Proc. I. R. E., **8**, p. 261; 1915.
134. Operating Features of the Audion, Amplification, etc.; E. H. Armstrong; Elec. World, **64**, p. 1149; 1914. Proc. I. R. E., **8**, p. 215, 1915; **5**, p. 145, 1917.
135. Characteristic Curves, and Use as Source of High Frequency Current; J. Bethenod; La Lumière Electrique, **35**, pp. 25, 225; 1916.
136. Generalized Equations for Audions; M. Latour; La Lumière Electrique, Dec. 30, 1916. Electrician, **78**, p. 280; 1916.
137. Characteristics of Audion Tubes Used in Radiotelegraphy; G. Vallauri; L'Elettrotecnica, **4**, Nos. 3, 4, 18, and 19; 1917.
138. Use of Pliotron to Produce Extreme Frequencies, Currents, and Voltages; W. C. White; General Electric Review, **19**, p. 771, 1916; **20**, p. 635, 1917.

MISCELLANEOUS SOURCES OF HIGH-FREQUENCY CURRENT.

141. Disturbing Short Waves in Buzzer Circuits; S. Loewe; Jahrb. d. drahtl. Tel., 6, p. 325; 1912.
142. Production of Undamped Oscillations; M. Wien; Jahrb. d. drahtl. Tel., 1, p. 474; 1908. Physikalische Zs., 11, p. 76; 1910.
143. Impulse Excitation Transmitter; E. W. Stone; Proc. I. R. E., 4, p. 233, 1916; 5, p. 133, 1917.
144. Frequency Multipliers; A. N. Goldsmith; Proc. I. R. E., 3, p. 55;1915. W. H. Eccles; Electrician, 72, p. 944; 1914.
145. High-Frequency Alternator of Induction Type; General Electric Review, 16, p. 16; 1913.
146. High-Frequency Alternator Employing Rotating Magnetic Fields; R. Goldschmidt; Electrician, 66, p. 744; 1911. T. R. Lyle; Electrician, 71, p. 1004; 1913.
147. Duddell Arc; W. Duddell; Jour. Röntgen Soc., 4, p. 1; 1907.
148. Arc generator for laboratory purposes; F. Kock, Phys. Zeitschr., 12, p. 124; 1911.
149. Impact excitation of undamped waves; E. L. Chaffee; Jahrb. d. drahtl. Tel. 7, p. 483; 1913. Proc. Amer. Ac. Arts & Sci., 47, No. 9; p. 267; 1911.

PUBLICATIONS OF THE BUREAU OF STANDARDS BEARING ON RADIO MEASUREMENTS

UNITS AND INSTRUMENTS

151. Units of Weight and Measure; Circular No. 47; 1914.
152. Electric Units and Standards; Circular No. 60; 1916. International System of Electric and Magnetic Units; J. H. Dellinger; Bull., 18, p. 599; 1916 (S. P. 292).
153. Electrical Measuring Instruments; Circular No. 20; 2d ed., 1915.
154. Fees for Electric, Magnetic, and Photometric Testing; Circular No. 6; 7th ed., 1916.

ELECTRICAL PROPERTIES OF MATERIALS

161. Copper Wire Tables; Circular No. 31; 3d ed., 1914.
162. Electric Wire and Cable Terminology; Circular No. 37; 2d ed., 1915.
163. Insulating Properties of Solid Dielectrics; H. L. Curtis; Bull., 11, p. 359; 1914 (S. P. 234).

CAPACITY AND INDUCTANCE

171. The Testing and Properties of Electric Condensers; Circular No. 36; 1912.
172. Formulas and Tables for the Calculation of Mutual and Self Inductance; Rosa and Grover; Bull., 8, p. 1; 1911 (S. P. 169).
173. Various papers on inductance calculations; see Circular No. 24, "Publications of the Bureau of Standards."
174. The Absolute Measurement of Capacity; Rosa and Grover; Bull., 1, p. 153; 1904 (S. P. 10).
175. Measurement of Inductance by Anderson's Method, Using Alternating Currents and a Vibration Galvanometer; Rosa and Grover; Bull., 1, p. 291; 1905 (S. P. 14).
176. The Simultaneous Measurement of the Capacity and Power Factor of Condensers; F. W. Grover, Bull., 3, p. 371; 1907 (S. P. 64).
177. Mica Condenser as Standards of Capacity; H. L. Curtis, Bull., 6, p. 431; 1910 (S. P. 137).
178. The Capacity and Phase Difference of Paraffined Paper Condensers as Functions of Temperature and Frequency; F. W. Grover; Bull., 7, p. 495; 1911 (S. P. 166).

179. The Measurement of the Inductances of Resistance Coils; Grover and Curtis; Bull., **8**, p. 455; 1911 (S. P. 175).
180. Resistance Coils for Alternating Current Work; Curtis and Grover; Bull., **8**, p. 495; 1911 (S. P. 177).
181. A Variable Self and Mutual Inductor; Brooks and Weaver; Bull., **13**, p. 569; 1916 (S. P. 290).

RADIO SUBJECTS

191. The Influence of Frequency Upon the Self-Inductance of Coils; J. G. Coffin; Bull., **2**, p. 275; 1906 (S. P. 37).
192. The Influence of Frequency on the Resistance and Inductance of Solenoidal Coils; L. Cohen; Bull., **4**, p. 161; 1907 (S. P. 76).
193. The Theory of Coupled Circuits; L. Cohen; Bull., **5**, p. 511; 1909 (S. P. 112).
194. Coupled Circuits in which the Secondary has Distributed Inductance and Capacity; L. Cohen; Bull., **6**, p. 247; 1909 (S. P. 126).
195. High-Frequency Ammeters; J. H. Dellinger; Bull., **10**, p. 91; 1913 (S. P. 206).
196. Direct-Reading Instrument for Measuring Logarithmic Decrement and Wave Length of Electromagnetic Waves; F. A. Kolster; Bull., **11**, p. 421; 1914 (S. P. 235).
197. Effect of Imperfect Dielectrics in Field of Radiotelegraphic Antennas; J. M. Miller; Bull., **13**, p. 129; 1916 (S. P. 269).

PUBLICATIONS OF THE UNITED STATES NAVAL RADIOTELEGRAPHIC LABARATORY IN THE BULLETIN OF THE BUREAU OF STANDARDS.

201. Detector for Small Alternating Currents and Electrical Waves; L. W. Austin; Bull., **1**, p. 435; 1905 (S. P. 22).
202. The Production of High-Frequency Oscillations from the Electric Arc; L. W. Austin; Bull., **3**, p. 325; 1907 (S. P. 60).
203. Some Contact Rectifiers of Electric Currents; L. W. Austin; Bull., **5**, p. 133; 1908 (S. P. 94).
204. A Method of Producing Feebly Damped High-Frequency Electrical Oscillations for Laboratory Measurements; L. W. Austin; Bull., **5**, p. 149; 1908 (S. P. 95).
205. The Comparative Sensitiveness of Some Common Detectors of Electrical Oscillations; L. W. Austin; Bull., **6**, p. 527; 1910 (S. P. 140).
206. The Measurement of Electric Oscillations in the Receiving Antenna; L. W. Austin; Bull., **7**, p. 295; 1911 (S. P. 157).
207. Some Experiments with Coupled High-Frequency Circuits; L. W. Austin; Bull., **7**, p. 301; 1911 (S. P. 158).
208. On the Advantages of a High Spark Frequency in Radiotelegraphy; L. W. Austin; Bull., **5**, p. 153; 1908 (S. P. 96).
209. Some Quantitative Experiments in Long Distance Radiotelegraphy; L. W. Austin; Bull. **7**, p. 315; 1911 (S. P. 159).
210. Antenna Resistance; L. W. Austin; Bull., **9**, p. 65; 1912 (S. P. 189).
211. The Energy Losses in Some Condensers Used in High-Frequency Circuits; L. W. Austin; Bull., **9**, p. 73 (S. P. 190).
212. Quantitative Experiments in Radiotelegraphic Transmission, L. W. Austin; Bull., **11**, p. 69; 1914 (S. P. 226).
213. Note on Resistance of Radiotelegraphic Antennas; L. W. Austin; Bull. **12**, p. 465; 1915 (S. P. 257).

APPENDIX 3.—SYMBOLS USED IN THIS CIRCULAR

B=magnetic induction.

c=velocity of light=2.9982×10^{10} cm per second.

C=electrostatic capacity.

d=diameter.

e=instantaneous electromotive force.

E=effective electromotive force.

E_0=maximum electromotive force.

$\mathscr{E}$=electric field intensity.

f=frequency.

F=force.

$\mathscr{F}$=magnetomotive force.

H=magnetic field intensity.

i=instantaneous current.

I=effective current.

I_0=maximum current.

$j=\sqrt{-1}$.

k=coupling coefficient.

K=dielectric constant.

l=length.

L=self-inductance.

m=mass.

M=mutual inductance.

p=instantaneous power.

P=average power.

Q=quantity of electricity.

r=distance from a point.

R=resistance.

$\mathcal{R}$=reluctance.

s=length along a path.

S=area.

t=time.

T=period of a complete oscillation.

v=velocity.

V=potential difference of a condenser.

w=instantaneous energy.

W=average energy.

X=reactance.

Z=impedance.

δ=logarithmic decrement.

ϵ=base of napierian logarithms = 2.71828.

θ=phase angle.

λ=wave length.

μ=permeability.

ρ=volume resistivity.

ϕ=magnetic flux.

ψ=phase difference.

$\omega=2\pi \times$frequency.

μf=microfarad.

$\mu\mu f$=micromicrofarad.

μh=microhenry.

Special symbols are defined where used in Part III and elsewhere.

DEPARTMENT OF COMMERCE

Circular

OF THE

Bureau of Standards

S. W. STRATTON, Director

No. 74

RADIO INSTRUMENTS AND MEASUREMENTS

ISSUED MARCH 23, 1918

PRICE, 60 CENTS

Sold only by the Superintendent of Documents, Government Printing Office
Washington, D. C.

WASHINGTON
GOVERNMENT PRINTING OFFICE
1918

This book may be kept

www.ingramcontent.com/pod-product-compliance
Lightning Source LLC
LaVergne TN
LVHW050151060326
832904LV00003B/112